Starry Nights: A Beginner's Journey Into Astronomy

Larry Culver

Published by Larry Culver, 2024.

While every precaution has been taken in the preparation of this book, the publisher assumes no responsibility for errors or omissions, or for damages resulting from the use of the information contained herein.

STARRY NIGHTS: A BEGINNER'S JOURNEY INTO ASTRONOMY

First edition. April 5, 2024.

Copyright © 2024 Larry Culver.

ISBN: 979-8224741830

Written by Larry Culver.

Preface

Embark on an Epic Journey and Empower Your Night Sky Observation Skills as a Beginner

In the vast expanse of the night sky, a cosmic spectacle unfolds, inviting us to embark on an extraordinary journey of discovery, as the stars beckon us to unravel the secrets they hold, and our adventure begins. "Starry Nights" is not just a guide; it is an invitation to witness the breathtaking beauty and profound mysteries that await new stargazers. Through this cosmic journey, we aspire to help and guide you, the beginning amateur astronomer, on a path that will engage you in a life-long endeavor to pursue the wonderful world of stargazing.

In the opening chapter, 'The Enchanting World of Stargazing', our cosmic exploration begins with an intriguing introduction that ignites curiosity and sets the tone for the adventure ahead. From the universal allure of stargazing to the emotional impact of the cosmos, the chapter unfolds as a celestial odyssey, where stars become storytellers and the night sky transforms.

In Chapter 2, "Telescopic Marvels - Unveiling the Cosmos", we help the beginning stargazer in selecting the right telescope with affordable options, define the essential telescope terminology and provide tips, resources, and accessories to enrich your stargazing experience.

In Chapter 3, "Stellar Stories - Tales of the Night Sky," we delve into ancient constellation tales. The beginning stargazer will learn to use star charts and mobile apps to enhance your understanding and identify celestial objects in the night sky - a key step in your astronomical journey.

"Ten Night Sky Sights for New Stargazers, takes the new stargazer on a tour of our recommendations for ten (10) celestial objects to observe and begin your celestial journey. From the Moon, constellations, planets, galaxies and meteor showers, these night sky objects are a great way to get started as a new stargazer, putting your newly found celestial map reading skills to work!

In the cosmic theater of Chapter 5, "Moonlit Marvels - Exploring Lunar Phases and Features", the moon assumes a central role, becoming a source of inspiration for human expression across art, literature, and the rhythmic ballet of its phases. Armed with a telescope, stargazers can delve into the lunar landscape, exploring craters, maria, and mountains, connecting with the cosmic narrative, illuminating our cosmic journey and evoking a sense of wonder.

In Chapter 6, "The Majestic Dance of the Planets", we explore the cosmic ballet of planets of our solar system, that have captivated humanity through the eons. This timeless dance reflects the enduring beauty and harmony of our celestial home.

"Deep Sky Odyssey - Exploring Nebulae, Galaxies, and Star Clusters" takes you on a cosmic journey beyond our solar system to uncover the hidden gems of deep space that have captivated humanity for centuries. This chapter inspires both beginners and seasoned astronomers to gaze into the cosmic wonders beckoning from the infinite expanse of deep space.

"Capturing the Night Sky" is an introduction to the beginner astronomer to world of astro-photography, as well as some tips for selecting the right camera and equipment for getting started.

Chapter 9, "Stargazing Etiquette and Safety", we discuss of the most important aspects of stargazing and astronomy - the appreciation and preservation of dark sky areas, as well as steps to take in protecting your eyes and equipment.

Chapter 10, "Resources for Further Exploration" identifies the wealth of resources available to enhance your learning and stargazing experience, providing you with a comprehensive list of recommended books and websites that will serve as valuable tools in your astronomical pursuits.

"Teaching Astronomy to Others" is for those wanting the rewarding experience of teaching astronomy to kids - it not only instills a love for science but also encourages curiosity about the universe. We give the teacher some tips and techniques for effectively teaching astronomy to kids, ensuring an engaging and educational experience.

STARRY NIGHTS: A BEGINNER'S JOURNEY INTO ASTRONOMY

I sincerely hope that your nights ahead are filled with wonder and cosmic discoveries, and that the beauty of the night sky continues to inspire. May each moment spent beneath the celestial sphere be a reminder of the vastness, mystery, and enduring beauty of the cosmos.

Table Of Contents

Preface

Chapter 1: The Enchanting World of Stargazing

The Night Sky Beckons: A Journey Begins

Universal Allure of Stargazing

Emotional Resonance of the Cosmos

Setting Realistic Expectations

Selecting the Perfect Stargazing Spot

Chapter 2: Telescopic Marvels - Unveiling the Cosmos

Choosing the Right Telescope

Understanding Telescope Terminology

Additional Tips and Resources

Essential Accessories for Stargazing

The Best 5 Beginner Telescopes Under $300

Chapter 3: Stellar Stories - Tales of the Night Sky

Understanding Constellations

Decoding the Night Sky's Map

Learning to Read the Sky

Mythic Narratives in the Stars

Indigenous Star Navigation Across Oceans

Modern Starlore - Constellations and Stories of Today

Steps for Developing Your Stargazing Skills

The Ecliptic: A Guiding Path of Celestial Wanderers

Using Stargazing Books, Star Charts and Apps

Top Apps for Stargazing and Astronomy

Additional Resources for Star Maps and Apps

Concluding Remarks

Chapter 4: Ten Night Sky Sights for New Stargazers

The Moon - Our Celestial Companion

Orion's Belt

Sirius - the Dog Star

The Pleiades - A Celestial Cluster of Seven Sisters

The Big Dipper

Jupiter: The King of Planets & Its Enchanting Moons

Mars – The Red Planet

The Andromeda Galaxy

The Summer Triangle

Meteor Showers and Celestial Fireworks

Chapter 5: Moonlit Marvels - Lunar Phases and Features

The Phases of the Moon

Lunar Features to Observe

Lunar Eclipses

Tips for Observing Lunar Eclipses

Equipment List for Observing a Lunar Eclipse

Tips for Lunar Observation

Earth's Tidal Dance - Lunar Influence on Tides

Human Footprints and Robotic Pioneers

Unexplained Lunar Anomalies & Mysteries

Earthrise and Cosmic Perspectives

Future Lunar Exploration - Beyond the Horizon

Chapter 6: The Majestic Dance of the Planets

The Planetary Symphony Begins

The Dazzling Inner-Planet Wanderers

Observing the Outer Planets

Planetary Alignments and Conjunctions

Cosmic Stage for the Planetary Dance

Concluding Remarks

Chapter 7: Deep Sky Odyssey: Nebulae, Galaxies, and Star Clusters

Introduction to Deep-Sky Objects

Star Clusters - Celestial Gatherings of Stellar Jewels

Observing Nebulae - Cosmic Clouds of Stardust

Galactic Marvels - Spirals, Ellipticals, and Irregulars

Cosmic Collisions and Galactic Dynamics

Cosmic Scale and the Limits of Exploration

Concluding Remarks

Chapter 8: Capturing the Night Sky

Introduction to Astrophotography

Choosing the Right Camera and Equipment

Tips for Beginner Astrophotographers

Chapter 9: Stargazing Etiquette and Safety

Respecting Dark Sky Areas

Protecting Your Eyes and Equipment

Dealing with Weather Conditions

Chapter 10: Resources for Further Exploration

Recommended Books and Websites

Joining Astronomy Clubs and Organizations

Attending Stargazing Events

Chapter 11: Teaching Astronomy to Others

Tips for Teaching Astronomy to Kids

Incorporating Astronomy into the Classroom

Inspiring the Next Generation of Stargazers

Conclusion: Embracing the Magic of the Night Sky

Starry Nights: A Beginner's Journey into Astronomy

Chapter 1: The Enchanting World of Stargazing

The Night Sky Beckons: A Journey Begins

Somewhere, something incredible is waiting to be known.

Carl Sagan and Sharon Begley (1977)

To begin our journey, let's understand what constellations are. Simply put, constellations are groups of stars that form recognizable patterns when viewed from Earth. These patterns often depict mythological characters, animals, or objects and serve as a guide to locate specific areas of the sky. It's important to note that constellations are not physically connected; they are merely a product of our perspective from Earth.

The history of constellations dates back thousands of years, with different cultures having their own set of recognized patterns. Ancient civilizations used constellations for various purposes, such as tracking seasons, determining planting times, or assisting in navigation. Today, we still use many of these ancient constellations, although their original meanings and stories may have evolved or been forgotten over time.

Identifying constellations can be a thrilling experience, especially for beginners. The first step is to familiarize yourself with the major constellations visible in your hemisphere. Start by locating prominent constellations like Orion, Ursa Major (the Big Dipper), or Scorpius. These easily recognizable patterns will serve as reference points for finding other constellations.

One helpful technique for identifying constellations is called star-hopping. This involves using easily identifiable stars or star patterns to navigate your way to the desired constellation. Begin with a bright star, and then follow a series of dimmer stars until you reach the target pattern. With practice, star-hopping becomes easier, and you'll be able to locate constellations effortlessly.

The night sky transforms into a celestial theater where constellations become the characters in a timeless play. From the graceful Orion to the mythical Ursa Major, each constellation tells a unique story that captivates the cosmic explorer. Understanding constellations also requires knowledge of their various components. Stars within a constellation may vary in brightness, color, and distance from Earth. Some constellations also contain deep-sky objects like star clusters, nebulae, and galaxies, which can be observed with binoculars or telescopes.

Let's unravel the incredible mysteries of the night sky!!

In the quiet solitude of the cosmic symphony, where stars emerge as notes in a celestial sonata, we, the Earthbound dreamers, find our gaze irresistibly drawn to the enchanting world of stargazing. The night sky, a canvas adorned with the brilliance of distant suns, becomes a portal inviting us to embark on a timeless journey—a journey that transcends the boundaries of our terrestrial existence.

Have you ever gazed up at the night sky and felt a sense of awe and wonder? The vast expanse of twinkling stars, the mysterious moon, and the occasional shooting star can captivate our imagination and ignite a curiosity about the universe. Welcome to our book, "Starry Nights: A Beginner's Journey into Astronomy." This book is specifically designed for new amateur astronomers, kids, and adults new to astronomy, entry-level telescope purchasers, and teachers.

We focus on providing tips, guides, and resources for those new to stargazing, as we explore the beauty and mysteries of the night sky, and equip you with the knowledge and tools to embark on your own astronomical journey.

We will introduce you to the celestial wonders that await you. From the constellations that have been captivating humans for centuries to the mesmerizing planets that grace our night sky, you will discover the extraordinary sights that can be observed with the naked eye or a basic telescope. We will also delve into the phases of the moon and explain how to identify and appreciate its changing appearance.

Next, we will guide you through the essential equipment needed for stargazing. Whether you are considering purchasing your first telescope or simply want to enhance your experience with binoculars, we will provide valuable advice on selecting the right equipment for your needs and budget. Additionally, we will highlight some useful accessories and resources that can enhance your stargazing adventures.

In our pursuit of exploring the night sky, we will discuss the importance of dark skies and offer tips on finding suitable stargazing locations. Light pollution can diminish the clarity and beauty of celestial objects, so we will show you how to escape the glow of city lights and experience the true magnificence of the stars.

To deepen your understanding and appreciation, we will introduce you to the basic concepts of astronomy. From the life cycles of stars to the phenomena of eclipses and meteor showers, you will gain insight into the scientific wonders that underpin the celestial ballet above us.

We will share some practical tips and techniques for observing the night sky. From understanding the importance of patience and adaptation to the night vision, to learning how to navigate the sky using star maps and smartphone apps, these valuable insights will help you make the most of your stargazing sessions.

Whether you are a child, adult, or teacher, the wonders of the night sky are accessible to all. So grab your telescope, or even just your curiosity, and join us on this beginner's journey into astronomy. Let the wonders of the night sky spark your imagination and ignite a lifelong passion for the vast universe that surrounds us. Happy stargazing!

Universal Allure of Stargazing

Gazing upon the celestial expanse is not a mere scientific pursuit; it is an instinctual, universal calling that has echoed through the corridors of human history. From the earliest civilizations who navigated by the stars to the modern dreamers peering through telescopes, the allure of stargazing transcends cultural and temporal confines. It beckons us to look beyond, to connect with the cosmos in a profound dance of curiosity and wonder.

Stargazing is a captivating and fulfilling hobby that offers a multitude of benefits to individuals of all ages. Whether you are a new amateur astronomer, a kid or an adult venturing into astronomy for the first time, or an entry-level telescope purchaser, exploring the wonders of the night sky can be a truly transformative experience. We will delve into the various benefits of stargazing, highlighting why it is an ideal pursuit for beginners.

Emotional Resonance of the Cosmos

The night sky is not a distant spectacle; it is a living, breathing entity that stirs profound emotions within us. As we gaze into the cosmic abyss, a sense of awe and humility envelops our consciousness. The realization of our minuscule presence in the vastness of space evokes a sublime mixture of introspection and inspiration. The stars, distant yet intimately connected, become conduits for our emotional journey into the cosmos.

As you gaze up at the celestial expanse, you can't help but feel a profound sense of humility and appreciation for the vastness and beauty of the universe. This sense of awe fosters a deeper connection with the natural world and can provide a much-needed respite from the stress and busyness of everyday life.

Stargazing encourages curiosity and a thirst for knowledge. As you begin to identify constellations, planets, and other celestial objects, you will naturally find yourself seeking to learn more about them. This curiosity can lead to a lifelong passion for astronomy and scientific exploration.

In addition to its intellectual and emotional benefits, stargazing also offers numerous practical advantages. For children, stargazing can be an excellent educational tool, teaching them about astronomy, physics, and the scientific method in a fun and engaging way. For adults, stargazing can provide a much-needed break from technology and screens, allowing for relaxation and mindfulness. Moreover, stargazing is a fantastic activity for families, couples, or friends to bond over, creating cherished memories and fostering a sense of togetherness.

For those considering purchasing their first telescope, stargazing offers the perfect opportunity to put your new equipment to use. By observing celestial objects through a telescope, you can unlock a whole new level of appreciation and understanding of the universe. This hands-on experience can be incredibly rewarding and will motivate you to continue exploring the depths of the night sky.

Stargazing can inspire individuals to become caretakers of the environment. Understanding the fragility and beauty of the universe can motivate us to preserve and protect our own planet. By fostering a sense of interconnectedness with the cosmos, stargazing can ignite a passion for environmental conservation and sustainability.

The allure of stargazing is vast and varied. From fostering a sense of wonder and curiosity to providing educational and practical advantages, stargazing is an ideal pursuit for beginners. So, grab your telescope, head outside, and embark on a journey into the vast expanse of the night sky – you won't be disappointed.

Setting Realistic Expectations

When diving into the fascinating world of astronomy, it's important for new amateur astronomers, both kids and adults, to set realistic expectations. Stargazing is an awe-inspiring hobby that can ignite a lifelong passion for the cosmos, but it's crucial to understand the limitations and challenges that come with it. In this subchapter, we will explore how to manage expectations and ensure that your journey into astronomy is both fulfilling and enjoyable.

First and foremost, it's important to acknowledge that astronomy is a complex field that requires time and dedication to truly grasp. As a beginner, it's unrealistic to expect to become an expert overnight. Instead, embrace the learning process and take small steps towards understanding the celestial wonders above.

One common misconception is that owning a telescope will instantly transport you to the depths of the universe. While telescopes can provide incredible views, they have their limitations. Entry-level telescope purchasers should understand that their instruments may not reveal the kind of detailed images seen in magazines or documentaries. However, with practice and patience, remarkable observations can still be made.

Teachers and parents introducing astronomy to children should emphasize that not every night will yield breathtaking views. Weather conditions, light pollution, and atmospheric disturbances can all impact what can be seen. Encourage young astronomers to be patient and appreciate the incremental improvements they make in their observations.

It's also important to manage expectations when it comes to identifying celestial objects. While the night sky is filled with countless wonders, it takes time to learn to distinguish between stars, planets, nebulae, and galaxies. Teaching beginners to start with the brightest and most easily recognizable objects, such as the Moon and major constellations, will help build confidence and enthusiasm.

Additionally, understanding the importance of planning and research is key to setting realistic expectations. Before heading out for a stargazing session, encourage beginners to check weather forecasts, moon phases, and light pollution levels. By being prepared, they can maximize their chances of having a successful and fulfilling night under the stars.

Setting realistic expectations is vital for new amateur astronomers, kids, and adults alike. Embracing the learning process, understanding the limitations of equipment, and appreciating the challenges of stargazing will ensure a rewarding journey into astronomy. By managing expectations and remaining patient, beginners can develop a deep appreciation for the wonders of the universe and embark on a lifelong passion for stargazing.

Selecting the Perfect Stargazing Spot

Embark on a cosmic adventure by picking the perfect spot to witness the night sky's magic. Go beyond the city lights and find yourself a place far from the bright hustle and bustle, where the beauty of the stars can truly shine. Think open fields, quiet hills, or a peaceful lakeside—somewhere where the sky isn't drowned out by the city's glow. Let the celestial show unfold in its full, dazzling splendor, offering a front-row seat to the wonders of the universe.

- Tips on Optimal Locations: The journey into the night sky begins with the choice of the perfect vantage point. Seek locations away from the suffocating glow of urban lights, where the celestial spectacle can reveal itself in all its luminous glory.
- Mitigating Light Pollution: Light pollution, the dimming of the stars due to artificial illumination, is an adversary to the stargazer. Selecting locations with minimal light pollution allows us to witness

the true brilliance of the cosmos.

- Creating a Comfortable Atmosphere: Stargazing is a contemplative endeavor that requires comfort. Bring blankets, reclining chairs, and a thermos of warm tea to create an environment conducive to cosmic musings.

The night sky becomes a boundless tapestry inviting us to unravel its secrets. The cosmos, with its universal allure, cultural tapestry, and emotional resonance, beckons us to embark on a journey that transcends the limits of our earthly existence.

Welcome to the timeless odyssey of stargazing, where the stars become our storytellers, and the night sky unfolds as a celestial epic awaiting our discovery.

Chapter 2: Telescopic Marvels - Unveiling the Cosmos

Peer through the lens of a telescope and witness the universe in unprecedented detail. Join me as we unravel the secrets of telescopic observation and explore the cosmos up close.

Telescopic exploration is not limited to professional astronomers. Backyard observatories and amateur astronomers contribute significantly to cosmic discoveries. This chapter delves into the thriving world of citizen science, where enthusiasts actively participate in unraveling the mysteries of the universe.

In the enchanting voyage through telescopic marvels, we traverse the cosmos, guided by the spirit of exploration. From the celestial revelations of Galileo to the cosmic sentinel, Hubble, telescopic exploration unveils the universe's secrets. Citizen scientists, armed with backyard observatories, contribute significantly, overcoming challenges like light pollution through observing campaigns.

When selecting a telescope, it's important to consider your specific needs and goals. Are you interested in observing planets, deep-sky objects, or both? Do you plan on traveling with your telescope or observing from a fixed location? These factors will help determine the type and size of telescope that best suits you.

For entry-level telescope purchasers, it is generally recommended to start with a beginner-friendly option, such as a refractor or a reflector telescope. Refractors are known for their simplicity and portability, making them ideal for beginners. Reflectors, on the other hand, offer a larger aperture at a more affordable price, allowing you to observe fainter objects.

In the vast expanse of the night sky, there are countless wonders waiting to be discovered. For new amateur astronomers, the journey into astronomy can be both exciting and overwhelming. One of the most crucial decisions to make is choosing the right telescope. Whether you are a kid or an adult, a beginner in astronomy, or a teacher looking to inspire young minds, this subchapter aims to guide you through this process.

The lunar canvas, adorned with craters and mountain ranges, tells tales of cosmic collisions. Telescopic lenses dance with Jupiter's dynamic atmosphere and Saturn's majestic rings. Mars, the red planet, showcases polar caps and atmospheric dynamics, captured through telescopic exploration. Beyond our solar system, telescopic marvels reveal the birth of stars in nebulae and the spiral arms of galaxies like Andromeda. Comets' cosmic tails and asteroid belt mysteries become celestial wanderers up close.

Telescopic marvels come with their set of challenges, from light pollution to atmospheric conditions. Observing campaigns, coordinated efforts by amateur and professional astronomers, aim to overcome these challenges and enhance our understanding of the cosmos.

Choosing the Right Telescope

Choosing the right telescope becomes a celestial quest, navigating between simplicity, aperture size, and mount stability. Essential telescope terminology unveils the language of the cosmos, connecting beginners and seasoned astronomers alike.

When selecting a telescope, it's important to consider your specific needs and goals. Are you interested in observing planets, deep-sky objects, or both? Do you plan on traveling with your telescope or observing from a fixed location? These factors will help determine the type and size of telescope that best suits you.

For entry-level telescope purchasers, it is generally recommended to start with a beginner-friendly option, such as a refractor or a reflector telescope. Refractors are known for their simplicity and portability, making them ideal for beginners. Reflectors, on the other hand, offer a larger aperture at a more affordable price, allowing you to observe fainter objects.

The cosmos beckons with affordable beginner telescopes under $300, from Explore Scientific's harmonious blend to Orion's simplicity. Celestron's

AstroMaster with equatorial mount offers planetary and deep-sky observation, while SkyWatcher's collapsible design balances portability and aperture size. Celestron's PowerSeeker stands as a budget-friendly entry, providing decent views of the moon and planets.

Aperture, or the diameter of the telescope's main lens or mirror, is a crucial factor to consider. Generally, a larger aperture allows more light to enter, resulting in brighter and clearer views. However, larger apertures also mean bulkier and more expensive telescopes. It's important to strike a balance between your budget and the desired level of observation.

Another consideration is the mount. A sturdy and stable mount is essential for smooth tracking and comfortable viewing. The two main types are altazimuth and equatorial mounts. Altazimuth mounts are simpler to use and move the telescope vertically and horizontally. Equatorial mounts, although more complex, allow for easier tracking of celestial objects as they follow the Earth's rotation.

For teachers and parents introducing astronomy to children, it is crucial to choose a telescope that is both user-friendly and durable. Look for telescopes specifically designed for kids, with features like easy assembly, lightweight construction, and educational resources to enhance their learning experience.

Choosing the right telescope can greatly enhance your stargazing journey. Consider your goals, budget, and requirements when selecting a telescope. Remember, the best telescope is the one that suits your needs and inspires you to explore the wonders of the universe. Happy stargazing!

Understanding Telescope Terminology

When embarking on your journey into astronomy, it is essential to familiarize yourself with the various terms associated with telescopes. This subchapter aims to provide new amateur astronomers, kids, and adults, as well as entry-level telescope purchasers, with a comprehensive understanding of telescope terminology. By grasping these concepts, you will be better equipped to explore the wonders of the night sky.

1. Aperture: The aperture refers to the diameter of the telescope's primary lens or mirror. It determines the amount of light the telescope can gather, with larger apertures allowing for brighter and more detailed views of celestial objects.

2. Focal Length: The focal length represents the distance between the telescope's primary lens or mirror and the point where the light converges to form an image. It determines the magnification and field of view offered by the telescope.

3. Eyepiece: The eyepiece is the lens you look through to observe celestial objects. It magnifies the image formed by the telescope, allowing you to see distant objects more clearly. Eyepieces come in various focal lengths, which affect the magnification level.

4. Mount: The mount is the support system for your telescope. There are two main types: the alt-azimuth mount, which moves up and down (altitude) and left and right (azimuth), and the equatorial mount, which aligns with the Earth's axis and allows for easier tracking of celestial objects.

5. Focal Ratio: The focal ratio is the ratio of the telescope's focal length to its aperture. It determines the brightness and field of view of the image. Low focal ratios provide wider views, while high focal ratios yield narrow, magnified views.

6. Collimation: Collimation refers to the alignment of the telescope's optics. Proper collimation ensures optimal performance and sharp images. It may require adjustment if the telescope becomes misaligned.

7. Field of View: The field of view is the extent of the sky visible through the telescope. It is typically measured in degrees and determines how much of the night sky you can observe at once.

By familiarizing yourself with these essential telescope terminologies, you will be better equipped to choose the right telescope for your needs, understand its capabilities, and navigate the vast expanse of the night sky. Whether you are a beginner stargazer or a seasoned astronomy enthusiast, this knowledge will serve as a solid foundation for your astronomical adventures. Remember, the universe is waiting to be explored, and your journey starts with understanding the tools that will guide you through the cosmos.

Additional Tips and Resources

To ensure a successful stargazing experience, we provide additional tips for maintaining and using telescopes effectively. Here are some additional resources:

- Maintenance and Repair: Check out websites like Cloudy Nights Telescope Reviews[1] and Astronomy Forum[2] for advice and discussions on telescope maintenance and repairs.
- Educational Resources: For beginning astronomy students, explore resources like NASA\'s Astronomy Picture of the Day[3] and Sky & Telescope's How-To Guides[4] to deepen your understanding of the cosmos.
- Enhancements for Viewing Experience: Consider additional equipment like eyepiece filters for specific celestial objects and software like Stellarium[5] for enhancing your celestial navigation and planning.
- Podcasts and Astronomy Clubs: Listen to podcasts like Star Talk Radio[6] for engaging discussions on astronomy. Join local astronomy clubs or online communities like Astronomy Connect[7] to connect with experienced astronomers and enthusiasts.

---

1. https://www.cloudynights.com

2. https://www.astronomyforum.net/

3. https://apod.nasa.gov/apod/astropix.html

4. https://skyandtelescope.org/astronomy-equipment/choosing-astronomy-equipment/telescopes/

5. https://stellarium.org

6. https://www.startalkradio.net

7. https://www.astronomyconnect.com

Essential Accessories for Stargazing

As you embark on your journey into the fascinating world of astronomy, it is important to equip yourself with some essential accessories to enhance your stargazing experience. These tools will not only make your observations more enjoyable but also help you navigate the night sky with ease. In this subchapter, we will discuss the must-have accessories for new amateur astronomers, kids and adults new to astronomy, entry-level telescope purchasers, and teachers.

By familiarizing yourself with these essential telescope terminologies, you will be better equipped to choose the right telescope for your needs, understand its capabilities, and navigate the vast expanse of the night sky. Whether you are a beginner stargazer or a seasoned astronomy enthusiast, this knowledge will serve as a solid foundation for your astronomical adventures. Remember, the universe is waiting to be explored, and your journey starts with understanding the tools that will guide you through the cosmos.

1. Sky Map or Star Chart: A sky map is an indispensable tool for any stargazer. It helps you identify constellations, stars, and planets in the night sky. There are numerous smartphone apps available that utilize your device's GPS to show you a real-time map of the stars above you.

2. Red LED Flashlight: A red LED flashlight is essential for preserving your night vision. Unlike white light, which can disrupt your eyes' adaptation to darkness, red light does not interfere with your ability to see celestial objects. This accessory is particularly useful when you need to consult star charts or adjust your equipment in the dark.

3. Binoculars: Binoculars are an excellent addition to your stargazing toolkit, especially for beginners. They provide a wider field of view compared to telescopes, allowing you to observe large celestial objects such as the Moon, star clusters, and even some galaxies. Look for binoculars with a high magnification power and a large objective lens diameter for optimal viewing.

4. Tripod: If you own a pair of binoculars or a small telescope, investing in a tripod will greatly stabilize your views. This accessory eliminates shaky hands,

providing a steady platform to observe celestial objects. A tripod will also allow you to free up your hands for sketching or taking notes.

5. Astronomy Apps and Software: There are several astronomy apps and software available that can enhance your stargazing experience. These tools provide real-time information on celestial events, satellite passes, and even offer virtual tours of the night sky. Some popular apps include SkySafari[8], Stellarium[9], and NASA's SkyView[10].

By equipping yourself with these essential accessories, you will be well-prepared to embark on your stargazing journey. Whether you are a beginner or an experienced observer, these tools will help you navigate the vast expanse of the night sky and discover the wonders it holds. So gather your gear, find a dark spot away from city lights, and let the stars guide you on a captivating astronomical adventure.

The Best 5 Beginner Telescopes Under $300

Selecting the right telescope as a beginner is crucial for an enjoyable and fulfilling stargazing experience. In this guide, we'll explore telescopes under $300, catering to beginners seeking quality optics on a budget.

We have provided a structured approach to presenting the telescope comparison, explaining the rationale behind each ranking, and offering recommendations based on user preferences and needs.

The telescopes are listed in order of our ranking, with the "best" telescope under $300 listed first. Our recommendations in this class of astronomical telescopes for beginners under $300 are:

1. Explore Scientific FirstLight 102 mm

2. Orion SkyScanner 100

---

8. https://skysafariastronomy.com/

9. https://stellarium.org/

10. https://skyview.gsfc.nasa.gov/current/cgi/titlepage.pl

3. Celestron Astromaster 114 EQ

4. SkyWatcher Heritage 130P Flextube

5. Celestron PowerSeeker 70

Please note that rankings are subjective and can vary based on individual preferences and requirements. Always consider your specific needs and preferences when choosing a telescope.

The optical quality and design specifications are provided based on available information, and it's always recommended to check the manufacturer's specifications for the most accurate details.

1. Explore Scientific FirstLight 102 mm[11]

Manufactured by Explore Scientific, a renowned name in producing high-quality optical instruments, the Explore Scientific FirstLight 102mm is a refractor telescope celebrated for its apochromatic lens, ensuring superior image quality. Paired with the Twilight I Mount, this telescope combines stability and ease of use.

---

11. https://explorescientificusa.com/products/fl-ar102600tn

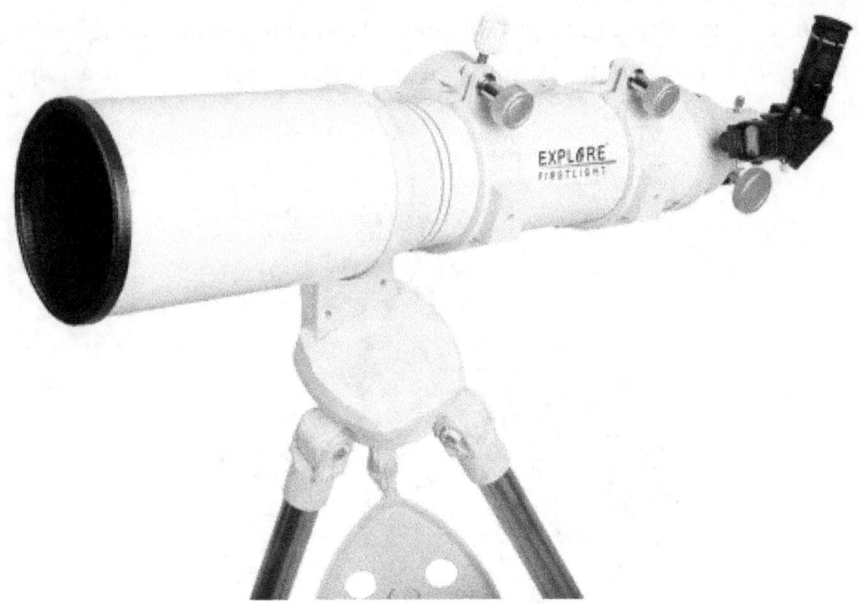

Estimated Price: $299.99

The Explore Scientific FirstLight 102mm Doublet Refractor is a telescope that is suitable for beginners. It has a 102mm aperture and a focal ratio of f/6.5. It is also camera adaptable and has an altazimuth mount.

This telescope clinches the top spot in our rankings due to its well-balanced features, including a moderate aperture, excellent optical quality, and a versatile design. It's an ideal choice for those who prioritize a harmonious blend of portability and image clarity. The telescope is recommended for individuals interested in observing the moon, stars, planets, as well as nebula/galaxies.

With an estimated price of $299.99, the Explore Scientific First Light 102mm[12] would be a great first choice for a beginner stargazer.

2. Orion SkyScanner 100[13]

---

12. https://www.stargazerspost.com/%22https://explorescientificusa.com/products/ fl-ar102600tn?_pos=1&_sid=afbb311fa&_ss=r/%22

13. https://www.telescope.com/Orion/Orion-SkyScanner-100mm-TableTop-Reflector-Telescope/rc/2160/p/ 102007.uts?keyword=orion%20skyscanner%20100

Crafted by Orion Telescopes & Binoculars, the Orion SkyScanner 100 is a reflector telescope featuring a tabletop Dobsonian mount, designed with simplicity and affordability in mind. Its compact design makes it a popular choice among beginners.

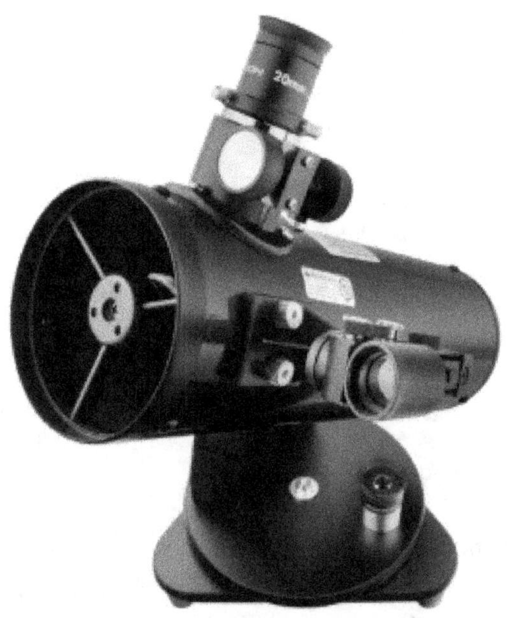

Estimated Price: $149.99

The Orion SkyScanner 100mm is a tabletop reflector telescope that is suitable for beginners. It has a 100mm parabolic glass primary mirror with a focal length of 400mm, a 1.25″ rack-and-pinion focuser, and two Kellner-type eyepieces. It also comes with an EZ Finder II red dot finder and an adjustable field tripod.

This telescope secures a high ranking due to its user-friendly design, simplicity, and cost-effectiveness. The SkyScanner 100 is recommended for entry-level astronomers seeking an easy-to-use and budget-friendly option, particularly for observing the moon and planets. Collimation is a trial and error process, primary mirror is difficult to fix if it gets knocked out of alignment, and it\'s best to supervise children while using it.

3. Celestron Astromaster 114 EQ

Even though this one may come at a price point right around $300, we feel that the decent aperture makes it very suitable for planetary and deep-sky observation.

Estimated Price: $290.00

The Celestron AstroMaster 114EQ is a Newtonian reflector telescope with an aperture of 114 mm and a focal length of 1000 mm. It comes with two eyepieces, a 20mm and a 10mm, and has a built-in red dot finder. It also has a pre-assembled tripod for easy setup.

This telescope earns its ranking due to the equatorial mount, sturdy construction, and balanced features. Recommended for users looking to step beyond entry-level telescopes for a richer observing experience, particularly for both planetary and deep-sky observation.

4. SkyWatcher Heritage 130 Flextube

Manufactured by SkyWatcher, known for innovative and high-quality astronomical instruments, the SkyWatcher Heritage 130P Flextube is a collapsible tabletop reflector telescope with a moderate aperture. Its collapsible design adds to its appeal, making it easy to store and transport.

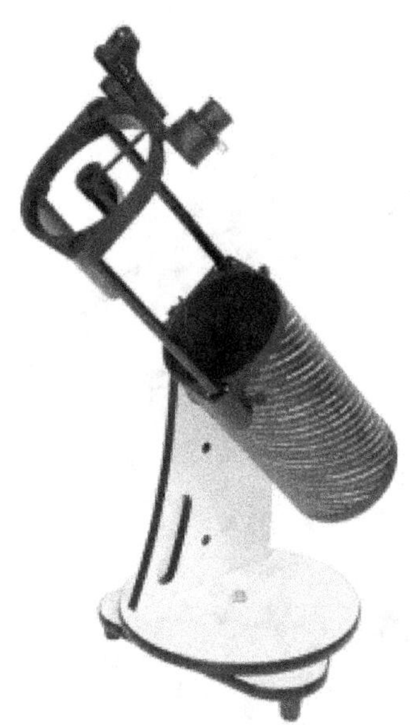

Estimated Price: $275.00

The Skywatcher Heritage 130P Flextube is a 130 mm parabolic dobsonian telescope with a 130 mm aperture. It has a collapsible tube and a retractable upper tube assembly that extends from 38 cm to 61 cm. The Heritage 130P comes with 10 mm and 25 mm 1.25 inch eyepieces with magnifications of 26x and 65x. It also includes a red dot finder.

This telescope's ranking is attributed to its compact design, collapsible structure, and a good balance between portability and aperture size. It is recommended for users valuing portability and seeking a telescope suitable for both planets and deep-sky objects.

5. Celestron PowerSeeker 70 Scope

Another creation by Celestron, the Celestron PowerSeeker 70 is an affordable refractor telescope with an altazimuth mount, designed for ease of use and accessibility. Its compact design and user-friendly features make it suitable for beginners.

Estimated Price: $124.95

The Celestron PowerSeeker 70 is a refractor telescope that is good for terrestrial and celestial viewing. It has a 70 mm aperture and a 700 mm focal length. It also has a German equatorial mount that allows for fine adjustments.

Earning its spot due to affordability, simplicity, and quick setup, this telescope is recommended for budget-conscious beginners seeking an easy entry into astronomy. An excellent choice for those on a budget, providing decent views of the moon and planets.

The choice of a beginner telescope involves considering personal preferences and intended use. Each of these telescopes, crafted by reputable manufacturers, offers a unique set of features catering to different needs and preferences.

Consider your priorities, and embark on your astronomical journey with confidence.

Chapter 3: Stellar Stories - Tales of the Night Sky

In the silent canvas of the night sky, stars tell stories that have echoed through millennia. Join me in uncovering the rich tapestry of stellar myths, legends, and cosmic tales.

In the cosmic ballet of the night sky, stars become storytellers, weaving tales that resonate through the ages. Constellations, those celestial patterns adorning the vast expanse, serve as cosmic guides, captivating human imagination for centuries. These configurations, born from our earthly perspective, have rich histories across diverse cultures, used for navigation, storytelling, and inspiration. Star-hopping, a technique for constellation identification, turns the night sky into a timeless play where each constellation plays a unique role.

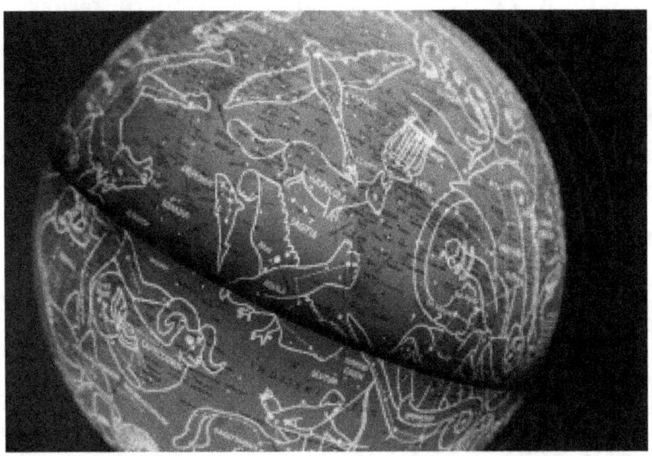

Navigating the celestial sphere has been revolutionized by star maps and astronomy apps. These tools, like "SkyView" and "Stellarium," transform the night sky into an interactive cosmic landscape. They not only assist in locating celestial objects but also provide real-time information and event notifications, enhancing the stargazing experience. Learning to read the sky involves mastering constellations, understanding celestial coordinates, and observing the motions of stars, planets, and the Moon.

For those taking their first steps in celestial exploration, star maps provide a splendid starting point, offering a tangible connection to the luminous wonders above. As you acquaint yourself with these maps, consider them as portals to the celestial realm, where constellations become stories written across the heavens. Begin by identifying the familiar patterns—the Orion's Belt, the Big Dipper, or the graceful sweep of Cassiopeia. These celestial landmarks serve as celestial beacons, anchoring your gaze and guiding you through the vastness of the cosmos.

Mythic narratives unfold in the stars, transcending cultures and time. From Greek mythology's Orion and Ursa Major to Aboriginal Australian Dreamtime constellations, the night sky is a canvas painted with tales of gods, heroes, and creatures. Indigenous cultures, like the Polynesians, navigated vast oceans using celestial guides, turning stars into celestial maps. Modern starlore incorporates human achievements in space, merging scientific discoveries with ancient cosmic stories.

As we journey through this exploration of stellar stories, the night sky reveals itself as a silent embrace of timeless sagas. Learning to read the sky is an ongoing adventure, where the stars become eternal teachers, inviting us to join the cosmic narrative unfolding in the silent night sky.

In the enchanting voyage through telescopic marvels, we traverse the cosmos, guided by the spirit of exploration. From the celestial revelations of Galileo to the cosmic sentinel, Hubble, telescopic exploration unveils the universe's secrets. Citizen scientists, armed with backyard observatories, contribute significantly, overcoming challenges like light pollution through observing campaigns.

The lunar canvas, adorned with craters and mountain ranges, tells tales of cosmic collisions. Telescopic lenses dance with Jupiter's dynamic atmosphere and Saturn's majestic rings. The moons of Jupiter, in cosmic harmony, reveal their unique characteristics. Io's volcanic drama and Enceladus' subsurface ocean offer glimpses into the dynamic forces shaping our solar system.

Mars, the red planet, showcases polar caps and atmospheric dynamics, captured through telescopic exploration. Beyond our solar system, telescopic marvels reveal the birth of stars in nebulae and the spiral arms of galaxies like Andromeda. Comets' cosmic tails and asteroid belt mysteries become celestial wanderers up close.

Understanding Constellations

One of the most fascinating aspects of the night sky is the presence of constellations. These celestial patterns have captured the imagination of humans for centuries and have been used as navigational tools, storytelling devices, and sources of inspiration. Next, we will delve into the world of constellations, exploring their history, significance, and how to identify them in the vast expanse of the night sky.

By grasping the significance and artistry of constellations, you'll unlock a whole new level of appreciation for the night sky. Whether you're a new amateur astronomer, a child or adult new to astronomy, an entry-level telescope purchaser, or a teacher looking to inspire young minds, understanding constellations is an essential foundation for stargazing. So grab your star charts, step outside, and let the wonders of the cosmos unfold before your eyes. Happy stargazing!

Decoding the Night Sky's Map

One of the most exciting aspects of stargazing is exploring the vastness of the night sky and discovering the wonders it holds. However, navigating through the celestial sphere can be challenging, especially for beginners. Luckily, with the advancements in technology, there are now star maps and apps available that can make this journey much easier and more enjoyable. Lets explore the benefits of using star maps and apps for stargazing and provide recommendations for the best ones to use.

Breaking down the night sky into manageable sections, celestial cartography becomes our guide through the cosmic landscape. Star maps have been used for centuries as a tool to locate and identify celestial objects. These maps depict the positions of stars, constellations, planets, and other celestial bodies in the night sky. They provide a visual representation of the stars and their relative positions, helping astronomers navigate and find specific objects. For beginners, star maps offer a great starting point to familiarize themselves with the night sky and begin identifying constellations.

Understanding the celestial coordinates—right ascension and declination—becomes the compass guiding us through the celestial sphere.

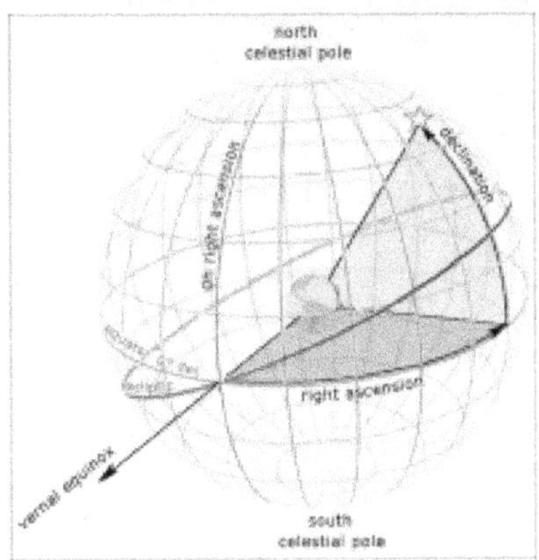

In addition to traditional star maps, there are now numerous astronomy apps available for smartphones and tablets. These apps utilize the device's GPS and motion sensors to provide real-time information about the night sky. They can accurately identify stars, planets, constellations, and even satellites based on the device's location and orientation. With just a few taps on the screen, users can access detailed information about the objects they are observing, such as their names, distances, and interesting facts.

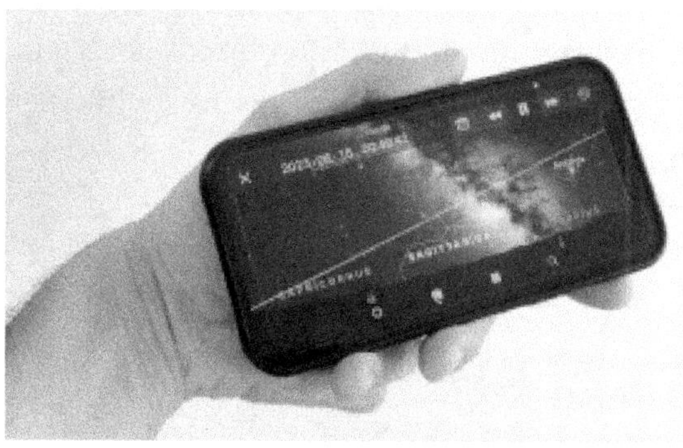

One of the advantages of using star maps and apps is their ability to assist in planning stargazing sessions. They can provide information about upcoming celestial events like meteor showers, eclipses, and planetary alignments. By knowing the precise times and locations of these events, amateur astronomers can optimize their observing sessions and increase their chances of witnessing spectacular phenomena.

For new amateur astronomers, kids, and adults just starting their journey into astronomy, star maps and apps are invaluable tools to enhance their stargazing experience. They offer a way to engage with and learn about the night sky in an interactive and accessible manner. Teachers can also utilize these resources to introduce astronomy concepts to students and make the learning process more engaging and immersive.

Some highly recommended star maps and apps for beginners include "SkyView", available on iOS and Android" "Star Walk," and "Stellarium." These apps are user-friendly, offer a wealth of information, and have intuitive interfaces that make them suitable for all ages and experience levels. Whether you are using a traditional star map or a smartphone app, incorporating these tools into your stargazing adventures will undoubtedly enhance your understanding and appreciation of the night sky.

Learning to Read the Sky

In the vast expanse of the night sky, there lies a world waiting to be explored. For new amateur astronomers, kids and adults alike, delving into the wonders of astronomy can be an exhilarating journey. However, before embarking on this adventure, it is essential to learn to read the sky, for it holds the key to unlocking the secrets of the universe.

The night sky is a canvas, adorned with countless stars, planets, and celestial objects. Learning to navigate this celestial map is the first step towards becoming a skilled stargazer. Understanding the basic tools and techniques for reading the sky will enable you to appreciate the beauty and magnitude of what lies above.

One of the fundamental skills in reading the sky is identifying constellations. These patterns of stars have captivated humans for centuries, each telling a unique story. By recognizing prominent constellations, such as Orion the Hunter or Ursa Major, you can begin to navigate the celestial sphere with confidence.

Another crucial aspect of reading the sky is understanding the motions of celestial objects. The apparent movement of stars throughout the night is due to the Earth's rotation. Observing this motion will help you determine the time of night and locate specific objects. Additionally, tracking the path of the Moon and planets can be a fascinating endeavor, as they traverse the sky in their own unique patterns.

Moreover, learning to read the sky involves mastering the art of using star charts and finding one's way around the night sky. Star charts are like roadmaps to the stars, providing a visual representation of the sky at different times of the year. By aligning these charts with your location and time, you can identify constellations, stars, and deep-sky objects.

Stellar Magnitudes

In the language of astronomers, the Greek alphabet takes center stage, offering designations to the brightness of stars through the stellar magnitude scale. This historical system, tracing its roots across the annals of astronomy, assigns lower magnitudes to brighter stars and higher magnitudes to fainter ones. It's akin to a celestial brightness code—a way for astronomers to quantify and compare the luminosity of stars. As you explore star maps, pay attention to these magnitudes. Witness how the ancient Greek alphabet seamlessly weaves into the cosmic narrative, allowing you to decipher the brilliance of stars like cosmic hieroglyphs.

While technology has ushered in precision and detailed measurements, the enduring use of the Greek alphabet-based stellar magnitude scale echoes the historical resonance of our cosmic quest. The stars, with their varied luminosities, beckon us to unravel their secrets, and star maps act as our celestial guides in this timeless pursuit. In essence, these maps transcend mere astronomical tools—they become gateways to the wonders of the universe, inviting you to partake in the grand cosmic narrative. So, unfold your star map, let it be your compass through the celestial seas, and may the stories of the night sky unveil themselves to you in all their cosmic splendor.

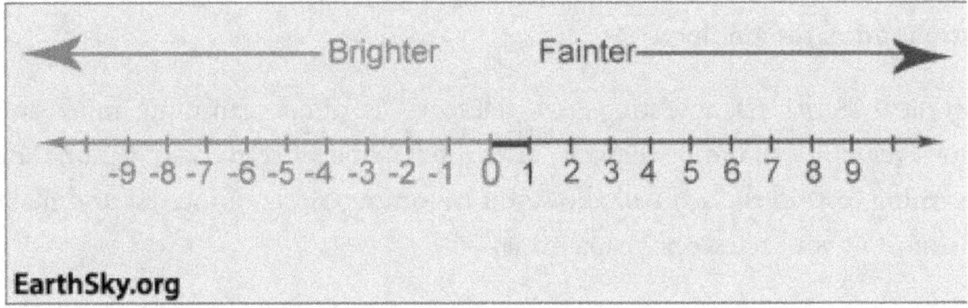

Here's a brief explanation of how the Greek alphabet letters are used in the stellar magnitude scale:

- Magnitude 1 (Brightest): The brightest stars in the night sky are assigned a magnitude of 1. Examples include Sirius (α Canis Majoris) and Vega (α Lyrae).

- Magnitude 2: Stars slightly fainter than the brightest are assigned a magnitude of 2. Examples include Canopus (α Carinae) and Altair (α Aquilae).

- Magnitude 3: The pattern continues, with decreasing brightness corresponding to increasing magnitude numbers. Examples include Aldebaran (α Tauri) and Spica (α Virginis).

- Magnitude 4: Stars of moderate brightness are assigned a magnitude of 4. Examples include Deneb (α Cygni) and Regulus (α Leonis).

- Magnitude 5: The faintest stars visible to the naked eye under optimal conditions have a magnitude of 5. Examples include Beta Centauri (β Centauri) and Epsilon Eridani (ε Eridani).

For very faint objects, numbers are sometimes used instead of Greek letters. The system allows astronomers to quantify and compare the brightness of celestial objects, providing a standardized way to express their luminosity. Modern technology and precise measurements have expanded our understanding, but the Greek alphabet-based stellar magnitude scale remains a part of astronomical terminology.

For new stargazers, investing in a telescope is often a thrilling milestone. However, understanding how to use a telescope effectively is paramount. Learning to read the sky will allow you to locate objects of interest and make the most of your telescope's capabilities.

Teachers, too, can benefit from learning to read the sky, as it enhances their ability to inspire and educate young minds. By sharing the knowledge of constellations, celestial motions, and star charts, teachers can ignite a passion for astronomy in their students, fostering a lifelong love for the cosmos.

Learning to read the sky is a journey that never truly ends. As you delve deeper into the world of stargazing, the sky becomes your companion, guiding you through the mysteries of the universe. So, embrace this adventure, and let the stars be your teachers.

Mythic Narratives in the Stars

Mythologies from different regions—Greek, Native American, and others—interweave tales of this celestial bear, marking it as a constant companion in the cosmic stories told by diverse cultures. Greek mythology intertwines with the constellations, giving rise to the Zodiac—a celestial belt where the epic stories of gods and heroes unfold. Across various mythologies, the dragon emerges as a celestial creature gracing the night sky. From Chinese constellations to European tales, the cosmic dragon becomes a mythical entity breathing fire into the stellar narratives of diverse cultures.

- From the mighty Hercules to the graceful Perseus, the night sky becomes a cosmic stage for the tales of ancient Greece. In Hindu mythology, the night sky becomes a canvas painted with stories of gods, goddesses, and celestial battles.

- The epic tale of the Mahabharata finds its reflection in the stars, with constellations embodying the characters and events of this ancient Indian narrative. Native American cultures, such as the Navajo, weave intricate stories into the night sky. The Coyote, a prominent figure in Navajo mythology, dances across the celestial plains, and the stars become storytellers preserving the wisdom and traditions of indigenous peoples.

- In Aboriginal Australian cultures, Dreamtime constellations narrate the creation stories and ancestral journeys. The emu in the sky and other celestial beings populate the night sky, connecting the Aboriginal people with their cosmic heritage.

- In Japanese folklore, the stars Vega and Altair represent Orihime and

Hikoboshi, lovers separated by the Milky Way. The annual Tanabata festival celebrates their reunion, and the cosmic love story becomes a cultural celebration under the celestial canopy.

- Greek mythology introduces us to the story of Cygnus, the swan, and Lyra, the lyre, entwined in cosmic love. The tragic tale of transformation and celestial alignment becomes a timeless narrative painted in the night sky.

- The constellation Aries, associated with the ram, intertwines with the Greek myth of the Golden Fleece. The cosmic quest of Jason and the Argonauts, seeking the elusive fleece, becomes a celestial journey embedded in the stars.

- In Mesopotamian mythology, the constellation Taurus aligns with the epic of Gilgamesh. The cosmic adventures of this legendary hero, facing celestial challenges and seeking immortality, unfold in the night sky.

- Greek mythology introduces Pegasus, the winged horse, galloping across the heavens. The constellation becomes a symbol of cosmic inspiration and heroic endeavors, carrying mythical riders on celestial journeys.

Indigenous Star Navigation Across Oceans

Indigenous cultures, like the Polynesians, mastered the art of wayfinding using the stars. The cosmic stories in the Polynesian night sky served as celestial maps, guiding navigators across vast oceans and connecting distant islands through the wisdom written in the stars. The Inuit people, residing in the Arctic regions, wove star tales into their cultural fabric. Constellations like the Great Bear and the Northern Lights became celestial guides and storytellers in the Arctic night, preserving the traditions of Inuit communities.

Modern Starlore - Constellations and Stories of Today

In the modern era, humanity's exploration of space introduces new constellations formed by satellites and the International Space Station and the SpaceX Starlink satellite internet system. The stories of human achievements in space become part of the contemporary starlore, with cosmic tales etched by human ingenuity in the celestial sphere.

Scientific advancements in astronomy reveal celestial phenomena and structures that inspire awe. From black holes to pulsars, these cosmic discoveries become part of the ever-evolving stories told by scientists studying the stars.

Steps for Developing Your Stargazing Skills

1. Step 1: Familiarize Yourself with the Night Sky - Begin by observing the night sky regularly. Find a location away from city lights where

the sky is dark, providing optimal visibility. Spend time simply looking up and gaining familiarity with the overall patterns of stars. Take note of any consistently visible celestial objects.

2. Step 2: Learn the Major Constellations - Start with the major constellations that are prominent throughout the year, such as Orion, Ursa Major (containing the Big Dipper), and Cassiopeia. Identify these constellations by their distinctive shapes, and understand their positions in relation to each other.

3. Step 3: Utilize Star Charts or Apps - Incorporate star charts or astronomy apps to aid your learning. Choose beginner-friendly resources that provide visual representations of the night sky. Many charts include constellation lines and labels to guide your identification process.

4. Step 4: Focus on Seasonal Constellations - As the seasons change, new constellations become visible. Learn to recognize the patterns that dominate each season. This approach gradually expands your knowledge base and allows you to navigate the night sky year-round.

5. Step 5: Understand Celestial Coordinates - Introduce yourself to celestial coordinates such as right ascension and declination. While this may sound complex, understanding the basics will help you locate specific objects in the night sky using astronomical coordinates.

6. Step 6: Study the Moon's Phases - The Moon is a readily visible celestial object that undergoes distinct phases. Observe and understand the various phases of the Moon as it orbits Earth. Recognizing these phases will enhance your overall comprehension of celestial mechanics.

7. Step 7: Explore Planets and Bright Stars - Identify and track the movement of planets visible to the naked eye. Understand their apparent motion against the background stars. Bright stars, such as Sirius and Betelgeuse, also serve as excellent reference points in the night sky.

8. Step 8: Observe Meteor Showers and Celestial Events - Participate in meteor shower observations and keep track of significant celestial events. Witnessing these events not only adds excitement to your

stargazing experience but also deepens your understanding of celestial occurrences.

9. Step 9: Learn about Deep-Sky Objects - Gradually move beyond stars and planets to explore deep-sky objects like nebulae, galaxies, and star clusters. Understanding how these objects are distributed across the night sky contributes to a more comprehensive view of our cosmic neighborhood.

1. Step 10: Engage in Stargazing Communities - Connect with fellow stargazers and join astronomy clubs or online forums. Engaging in discussions, asking questions, and sharing your observations with others can provide valuable insights and create a supportive community for learning.

2. Bonus Tip: Keep a Stargazing Journal - Maintain a stargazing journal to document your observations, noting the date, time, and location. Record your impressions, identify new constellations, and track your progress. This personalized record serves as both a learning tool and a delightful retrospective of your astronomical journey.

The Ecliptic: A Guiding Path of Celestial Wanderers

The ecliptic is a fundamental concept in astronomy, serving as the apparent path that the Sun appears to trace across the celestial sphere throughout the year. It is the plane of Earth's orbit projected onto the sky, creating a visible line that intersects with the celestial sphere at an angle. For the beginning stargazer, understanding the ecliptic provides a key reference for locating planets, the Moon, and the Sun in the night sky.

The ecliptic is also central to observing celestial events such as solar and lunar eclipses. Eclipses occur when the Sun, Earth, and Moon align along the ecliptic plane. Solar eclipses transpire when the Moon passes between the Sun and Earth, casting a shadow on our planet. Lunar eclipses unfold when the Earth passes between the Sun and the Moon, causing the Moon to move into the Earth's shadow.

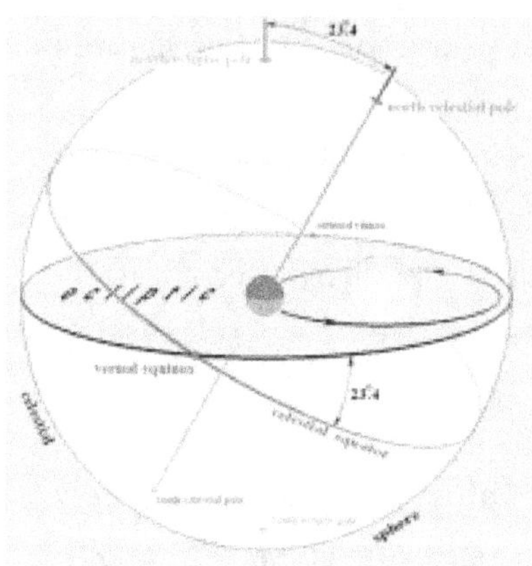

The Sun's apparent motion along the ecliptic influences the changing seasons on Earth. During the vernal equinox, the Sun crosses the celestial equator, marking the beginning of spring in the northern hemisphere. Conversely, the autumnal equinox occurs when the Sun crosses the celestial equator, ushering in fall. The summer solstice, when the Sun reaches its northernmost point along the ecliptic, and the winter solstice, when it reaches the southernmost point, delineate the peak of summer and the depths of winter, respectively.

The ecliptic cuts through various constellations, and the zodiacal constellations specifically align with this path. The zodiac consists of twelve constellations, each associated with a particular month and linked to the Sun's apparent path through the sky. The well-known zodiacal constellations include Aries, Taurus, Gemini, Cancer, Leo, Virgo, Libra, Scorpius, Sagittarius, Capricornus, Aquarius, and Pisces. These constellations serve as markers along the ecliptic, aiding stargazers in locating celestial objects.

Understanding the celestial poles is crucial in comprehending the ecliptic's orientation. The North Celestial Pole is the point in the sky around which all stars appear to rotate in the northern hemisphere, while the South Celestial Pole serves the same purpose in the southern hemisphere. The celestial poles are perpendicular to the plane of the Earth's rotation and are not fixed in the sky; they align with the Earth's axis of rotation. The angle between the celestial poles and the ecliptic is equal to the axial tilt of Earth, approximately 23.5 degrees.

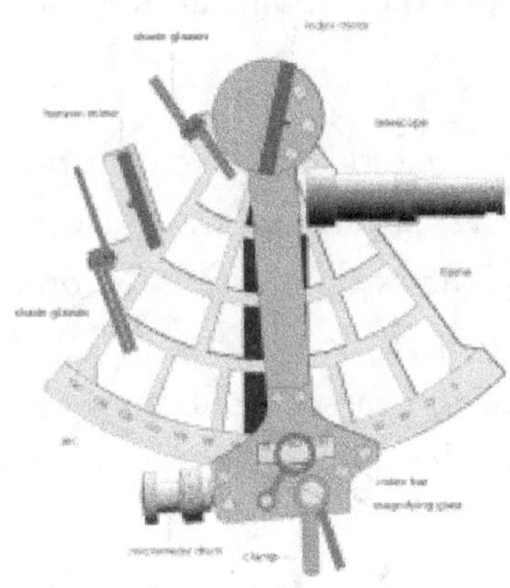

The ecliptic uses right ascension (RA) and declination (Dec) to precisely locate objects in the night sky. Right ascension is measured in hours, minutes, and seconds, analogous to the Earth's rotation, while declination is measured in degrees north or south of the celestial equator. The ecliptic intersects the celestial equator at two points: the vernal equinox (where the Sun crosses from the southern to the northern hemisphere) and the autumnal equinox (the reverse).

The tilt of the ecliptic, known as ecliptic obliquity, is a result of the Earth's axial tilt. This angle of inclination, approximately 23.5 degrees, contributes to the changing seasons and the Sun's varying position in the sky throughout the year. Ecliptic obliquity also influences the lengths of day and night at different times of the year.

For the beginner astronomer, the ecliptic is a valuable guide for locating planets and the Moon. Planets in our solar system primarily follow the plane of the ecliptic in their orbits. When observing the night sky, planets and the Moon are most frequently found along or near the ecliptic, aiding stargazers in identifying and tracking these celestial wanderers.

The ecliptic is central to observing celestial events such as solar and lunar eclipses. Eclipses occur when the Sun, Earth, and Moon align along the ecliptic plane. Solar eclipses transpire when the Moon passes between the Sun and Earth, casting a shadow on our planet. Lunar eclipses unfold when the Earth passes between the Sun and the Moon, causing the Moon to move into the Earth's shadow.

In summary, the ecliptic is a celestial guide that shapes the apparent path of the Sun, influences the changing seasons, and serves as a reference point for locating celestial objects in the night sky. As a foundational concept for beginning stargazers, the ecliptic enhances the understanding of the celestial coordinate system, the celestial poles, and the constellations along its course, fostering a deeper appreciation for the dynamic interplay of celestial phenomena.

Using Stargazing Books, Star Charts and Apps

One of the most exciting aspects of stargazing is exploring the vastness of the night sky and discovering the wonders it holds. However, navigating through the celestial sphere can be challenging, especially for beginners. Luckily, with the advancements in technology, there are now star maps and apps available that can make this journey much easier and more enjoyable. In this subchapter, we will explore the benefits of using star maps and apps for stargazing and provide recommendations for the best ones to use.

Learning the constellations can be an exciting journey, and the best method often depends on individual preferences. Here's a comparison of stargazing books, charts, projectors, and mobile applications to help you choose the method that suits you best:

Stargazing Books

Stargazing books serve as invaluable resources for both novice and experienced astronomers, offering a wealth of information and insights into the wonders of the night sky. One of their notable advantages lies in the in-depth exploration of the mythology and history associated with constellations. These books often provide captivating narratives that breathe life into the celestial patterns, connecting the dots between stars and cultural stories. By delving into the rich tapestry of myths, readers gain a deeper appreciation for the cultural significance of the night sky.

A significant benefit of stargazing books is the inclusion of detailed star maps and images. These visual aids play a crucial role in aiding observers to identify constellations and celestial objects with confidence. The intricate illustrations help bridge the gap between theoretical knowledge and practical observation, enabling stargazers to navigate the vastness of the night sky effectively.

However, it's essential to consider certain drawbacks when relying solely on stargazing books. One notable limitation is the static nature of the content. Unlike interactive tools or mobile applications, books lack the dynamic, real-time updates that align with the observer's location and time. This static quality may pose challenges for individuals seeking the most current information or those desiring interactive, hands-on learning experiences.

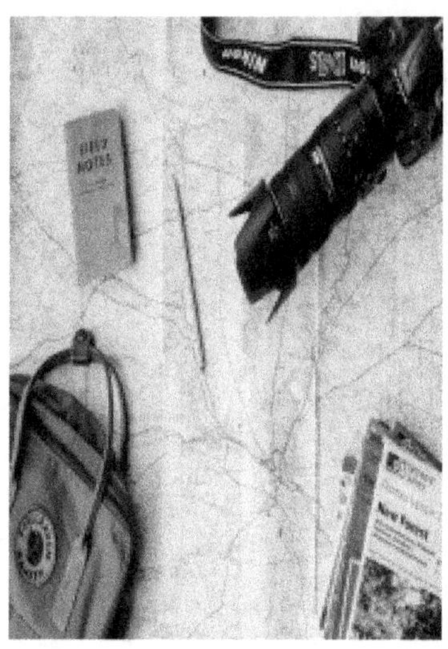

Additionally, stargazing books may be perceived as overwhelming for absolute beginners. The sheer volume of information, coupled with the intricate details provided, could potentially be intimidating for those taking their first steps into the realm of astronomy. In such cases, beginners might benefit from a more gradual introduction to stargazing, using a combination of resources tailored to their learning preferences and comfort levels.

For those enthusiastic about delving into stargazing through books, notable titles such as "NightWatch" by Terence Dickinson, "Turn Left at Orion" by Guy Consolmagno and Dan M. Davis, or "The Stars: A New Way to See Them" by H.A. Rey offer engaging content and user-friendly approaches. Supplementing book-based learning with other interactive tools can create a well-rounded and enriching stargazing experience for beginners, fostering a lifelong passion for astronomy.

Star Maps and Charts

For novice stargazers embarking on their celestial journey, star charts stand out as essential tools, offering both portability and ease of use in the field. The convenience of carrying a compact star chart, whether in physical or digital form, allows beginners to navigate the night sky with relative ease. The user-friendly nature of star charts makes them particularly advantageous for on-the-spot observation, providing a quick reference for identifying constellations and celestial objects.

Star maps[1] have been used for centuries as a tool to locate and identify celestial objects. These maps depict the positions of stars, constellations, planets, and other celestial bodies in the night sky. They provide a visual representation of the stars and their relative positions, helping astronomers navigate and find specific objects. For beginners, star maps offer a great starting point to familiarize themselves with the night sky and begin identifying constellations.

One notable advantage of star charts is their versatility in catering to specific time periods and locations. Many charts are tailored to cover precise dates, ensuring that stargazers have accurate and relevant information based on their observing times. This feature proves invaluable for those who wish to plan stargazing sessions in advance or track the positions of celestial bodies at specific moments, enhancing the overall experience of observing the night sky.

However, it's crucial to acknowledge certain challenges associated with using star charts, especially for beginners. Interpreting these charts can be initially daunting, as they often feature a multitude of symbols, lines, and labels. Novice astronomers may find it helpful to familiarize themselves with the basic elements of a star chart, such as cardinal directions and the concept of the celestial sphere, before venturing into more complex charts.

Another limitation of star charts lies in their inherent constraint to the information present on the chart itself. Unlike dynamic tools such as mobile applications or computer software, star charts provide a static snapshot of the night sky. This static quality means that users need to consult additional references or use complementary tools to access real-time information, planetary positions, or other dynamic aspects of the night sky.

For those eager to explore stargazing through star charts, resources such as "Sky & Telescope's Pocket Sky Atlas" by Roger W. Sinnott or "Stellarium," a popular and user-friendly planetarium software, offer accessible entry points. Beginners can gradually build their proficiency in interpreting star charts by starting with

---

1. https://www.space.com/star-charts.html

simpler versions and gradually progressing to more detailed and comprehensive charts as their understanding of the night sky deepens.

Astronomy Apps

For individuals stepping into the realm of astronomy or stargazing, mobile applications emerge as invaluable companions, offering a seamless blend of convenience and technology. The portability of these applications ensures that a wealth of astronomical information is constantly at the user's fingertips, transforming smartphones or tablets into personal observatories. This accessibility proves particularly advantageous for beginners eager to explore the night sky without the need for complex equipment.

One significant advantage of mobile applications is their ability to provide real-time sky tracking. These apps utilize built-in GPS functionality to precisely determine the observer's location and adjust the displayed sky accordingly. This dynamic feature ensures that stargazers receive accurate and current information about celestial objects visible from their specific geographic coordinates. As a result, users can confidently identify stars, planets, and constellations based on their real-time positions in the night sky.

STARRY NIGHTS: A BEGINNER'S JOURNEY INTO ASTRONOMY

Augmented reality (AR) features represent another compelling aspect of many astronomy apps, elevating the stargazing experience to new heights. Through AR technology, these applications overlay information about constellations, stars, and planets onto the live camera view of the device. This visual augmentation allows beginners to easily connect what they observe in the night sky with the corresponding celestial objects, facilitating a more immersive and educational stargazing experience.

One of the advantages of using star maps and apps is their ability to assist in planning stargazing sessions. They can provide information about upcoming celestial events like meteor showers, eclipses, and planetary alignments. By knowing the precise times and locations of these events, amateur astronomers can optimize their observing sessions and increase their chances of witnessing spectacular phenomena.

Despite these numerous benefits, it's essential for novice astronomers to be aware of certain considerations when using mobile applications. One limitation lies in the dependence on battery life, as extended stargazing sessions may deplete the device's power. To mitigate this, beginners can optimize their device settings, such as screen brightness and power-saving modes, to maximize battery longevity during observing sessions.

Additionally, some astronomy apps may require an internet connection for certain features, such as downloading updated celestial databases or accessing additional information. Users should consider their location and the availability of internet connectivity when choosing and using these applications.

For new amateur astronomers, kids, and adults just starting their journey into astronomy, star maps and apps are invaluable tools to enhance their stargazing experience. They offer a way to engage with and learn about the night sky in an interactive and accessible manner. Teachers can also utilize these resources to introduce astronomy concepts to students and make the learning process more engaging and immersive. The "Sky Live[2]" is a fully-interactive personal planetarium that you can tailor to your location and date/time.

---

2. https://theskylive.com/planetarium

Some highly recommended star maps and apps for beginners include "SkyView," "Star Walk," and "Stellarium." These apps are user-friendly, offer a wealth of information, and have intuitive interfaces that make them suitable for all ages and experience levels. Whether you are using a traditional star map or a smartphone app, incorporating these tools into your stargazing adventures will undoubtedly enhance your understanding and appreciation of the night sky. You can also browse the Star Gazer's blog article[3] for additional applications.

These user-friendly apps cater to beginners by offering intuitive interfaces, educational content, and interactive features that enhance the enjoyment of exploring the cosmos. Integrating these applications into one's stargazing toolkit can transform the learning experience and foster a lasting passion for the wonders of the night sky.

Top Apps for Stargazing and Astronomy

Tired of squinting at unfamiliar constellations and wishing you knew what that bright streak across the sky actually is? Ditch the guesswork and unlock the secrets of the cosmos with the power of your smartphone! Whether you\'re an iPhone enthusiast or an Android devotee, there\'s a universe of stargazing apps out there designed to transform your backyard into a personal planetarium. Dive into our curated list of hidden gems and rediscover the magic of looking up, ready to identify celestial wonders with ease.

There are so many things to see in the night sky, sometimes choosing one is quite a feat! You can leverage a few different mobile apps to narrow down what you would like to view and when. The following apps will help you save time by giving you key details about the moon, planets and stars that you can view in the universe.

SkyEye - The Best Free Stargazing App For Android

With the SkEye stargazing app, you get an amazing user experience for Android phones for free. Although there is a paid 'Pro' version, the amount of objects available in the free version is quite vast.

---

3. https://www.stargazerspost.com/2024/01/02/top-6-apps-for-stargazing-and-astronomy/

The nice thing about the targeted exploration is that once you select the object you would like to identify, the app tells you if you are close or far from seeing the object in your phone with a green, yellow or red indicator in the sky overlay.

I use this app almost exclusively when I'm trying to look at a specific object in my telescope. I simply search for the object I want to view, guide my phone by hand until the indicator circle is green and then line up the spotting scope on my telescope to match where my phone has found it, and I'm there!

The following features below make stargazing much easier for you:

- Vast Celestial Database: Explore stars, planets, nebulae, galaxies, and more.
- argeted Exploration: Search for specific objects by name or browse by type – stars, planets, nebulas, etc.
- Real-Time Sky Overlay: Point your phone, identify what you see.
- Definitely download SkyEye[4] for your Android and turn your phone into a powerful cosmic companion.
- ABSOLUTELY FREE

Nightsky - The Best Free Stargazing App For iPhone

The Nightsky app is a great choice if you are looking for free stargazing apps for iPhone. It really emphasizes ease of use and is a great app for beginners.

There are in-app purchases, but the features below come standard with the app straight from the App Store. If you have an iPhone and want a very easy-to-use app with the baseline features you want to get started, give Nightsky a try![5]

The features that make this app a worthy choice are:

- Simple interface: Quickly find the stars and constellations you\'re looking for with a clear and uncluttered interface.
- onight's highlights: See a curated list of the most visible celestial objects for your current location and time.

---

4. https://play.google.com/store/search?q=skeye&c=apps

5. https://apps.apple.com/us/app/night-sky/id475772902

- AR mode: Explore the sky in real-time with the Augmented Reality (AR) overlay.
- Constellation stories: Learn about the myths and legends behind the constellations.
- ABSOLUTELY FREE

The Best Paid Stargazing Apps For Android And iPhone

- SkySafari

Though a bit more on the pricey side for a mobile app ($9.99 in the Google Play Store), SkySafari 7 Plus is an excellent tool for tracking upcoming stargazing events. It has a calendar for all kinds of events from meteor showers, asteroids, planetary transits, etc. Even though SkySafari is fantastic to use as a sky map and for facilitating other stargazing activities, the mind-bogglingly huge database of astronomical phenomena alone makes it worth the price.

- Stellarium

While SkySafari 7 Plus undoubtedly shines as a celestial event maestro, there\'s another contender in the stargazing app arena that competes with it in many aspects: Stellarium Mobile.

Stellarium is a powerful, open-source tool that grants you unrivaled access to locations of celestial objects. The following are a list of awesome features:

- Enormous database – search over a million stars, hundreds of thousands of galaxies and nebulae, and even detailed simulations of planets and moons
- Vibrant community – Active forum of users and space enthusiasts
- 3D Sky Map
- Download Stellarium here![6]
- Celestial events calendar (with paid version)

---

6. https://stellarium.org/

Stellarium even lets you choose from a list of different sky cultures. From the default Western to Norse to even Tongan/Polynesian! Give Stellarium a try if you want a good all-around app for Android or iPhone. Even the non-paid version has plenty of great features to get started gazing!

So open Stellarium Mobile's free skies and explore a million stars, or let one of the other aforementioned apps be your guided tour through cosmic events. Whichever app sparks your curiosity, download your celestial companion and unlock the magic of the night sky!

Additional Resources for Star Maps and Apps

Whether you're delving into the cosmic mysteries online or seeking tangible resources offline, there are various tools and materials available to guide you in using star maps. Here are three recommendations for both online and offline resources:

Online Resources

- Sky & Telescope Interactive Sky Chart (Online):[7] Sky & Telescope's Interactive Sky Chart is an online tool that generates customizable star maps based on your location, date, and time. It helps users visualize the night sky, identify constellations, and locate celestial objects. The interactive features allow you to zoom in and out, making it a valuable resource for stargazers of all levels.
- Sky-Map.org:[8] Another great interactive sky map/sky chart. The main purpose of SKY-MAP is to consolidate astronomical, astrophysical and other information about different space objects and astrophysical facts.
- NASA's Night Sky Network (Online):[9] NASA's Night Sky Network

---

7. https://skyandtelescope.org/interactive-sky-chart/

8. http://www.sky-map.org/

9. https://nightsky.jpl.nasa.gov/

provides online resources for amateur astronomers and stargazers. It offers tools and articles on how to observe the night sky, including tips on using star maps. The website also features event calendars and educational materials for those interested in organizing or participating in astronomy-related activities.

Offline Resources

- "Turn Left at Orion" by Guy Consolmagno and Dan M. Davis (Book): "Turn Left at Orion" is a highly regarded book among amateur astronomers. It provides practical guidance on how to observe the night sky using binoculars or small telescopes. The book includes detailed star maps, making it an excellent offline resource for learning the art of stargazing. It covers a variety of celestial objects and offers clear instructions on how to locate them.
- Star Atlases (Printed Maps): Star atlases are printed maps of the night sky that you can use while observing outdoors. They often include detailed charts of constellations, stars, and deep-sky objects. Popular star atlases include "Sky Atlas 2000.0" and the "Cambridge Star Atlas." These physical atlases are handy tools for stargazers who prefer a tactile reference while exploring the night sky.
- Local Planetariums and Observatories: Many local planetariums and observatories offer in-person stargazing events and workshops. Check with your local astronomy community or educational institutions for scheduled programs. Attending these events can provide hands-on experience with using star maps under the guidance of experienced astronomers.
- Local Astronomy Clubs: Attending local astronomy club events or stargazing workshops can provide hands-on experience and guidance from experienced astronomers, making the learning process more interactive and enjoyable. Experiment with different methods to find what works best for you and enhances your understanding and appreciation of the night sky.

These online and offline resources cater to various learning preferences, ensuring that whether you prefer digital simulations or tangible materials, you can delve into the captivating realm of using star maps to navigate the celestial wonders above.

Concluding Remarks

For new stargazers, investing in a telescope is often a thrilling milestone. However, understanding how to use a telescope effectively is paramount. Learning to read the sky will allow you to locate objects of interest and make the most of your telescope's capabilities.

Teachers, too, can benefit from learning to read the sky, as it enhances their ability to inspire and educate young minds. By sharing the knowledge of constellations, celestial motions, and star charts, teachers can ignite a passion for astronomy in their students, fostering a lifelong love for the cosmos.

In this exploration of stellar stories, we have journeyed through the rich tapestry of myths, legends, and cosmic tales woven into the silent canvas of the night sky. From ancient constellations inspired by heroic journeys and love stories to the cultural wisdom embedded in indigenous starlore, the stars continue to narrate the timeless sagas of human imagination. As we gaze upon the celestial theater, the stories written in the stars transcend time and culture, inviting us to join the cosmic narrative that unfolds in the silent embrace of the night sky.

Learning to read the sky is a journey that never truly ends. As you delve deeper into the world of stargazing, the sky becomes your companion, guiding you through the mysteries of the universe. So, embrace this adventure, and let the stars be your teachers.

Chapter 4: Ten Night Sky Sights for New Stargazers

Discovering Ten Night Sky Sights as Your Guiding Constellations into the Cosmos for New Stargazers

For those embarking on their cosmic odyssey, it is paramount to commence with the contemplation of celestial entities that readily unveil themselves to our eager gazes, igniting an ardor that lays the groundwork for loftier celestial inquiries. Allow me to present an inventory of ten astral wonders adroitly suited for neophyte stargazers – a celestial primer, if you will, designed to kindle an intrinsic fascination with the cosmos and to beckon the mind towards loftier astronomical pursuits.

Embarking upon the celestial voyage, the nocturnal expanse unfolds as an enthralling canvas for those who turn their gaze skyward in earnest. Within this cosmic tapestry, I offer a compendium of ten astral phenomena, graciously bestowed upon the novice stargazer. These cosmic beacons, resplendent in their accessibility, beckon with an invitation to traverse the celestial realms and commence a journey of wonder and discovery.

1. The Moon: Observe the phases of the Moon and marvel at its craters and lunar landscapes.
2. Orion's Belt: Look for the three bright stars in a straight line, forming the iconic belt of the Orion constellation.
3. Sirius: Spot the brightest star in the night sky, located in the constellation Canis Major.
4. The Pleiades (Seven Sisters): A beautiful open star cluster that is easily visible to the naked eye.
5. The Big Dipper: Part of the Ursa Major constellation, this asterism is easily recognizable with its ladle-like shape.
6. Jupiter and It's Moons: Jupiter is often visible and its four largest moons, known as the Galilean moons, can be observed with binoculars.
7. Mars: Look for the reddish glow of Mars in the night sky, especially

during its opposition when it's closest to Earth.

8. The Andromeda Galaxy: The nearest spiral galaxy to our Milky Way, visible as a faint smudge under dark skies.
9. The Summer Triangle: Formed by the bright stars Vega, Deneb, and Altair, this prominent asterism is visible in the summer months.
10. Meteor Showers: Keep an eye out for meteor showers like the Perseids or Geminids for a dazzling display of shooting stars.

The Moon - Our Celestial Companion

Behold the preeminent luminary gracing our nightly tableau—the Moon, an ever-present celestial companion captivating in its prominence. Direct your gaze towards this ethereal wanderer, and with the aid of binoculars or a telescope, delve into an exploration of its phases and features. Through these optical instruments, the lunar landscape unveils itself with remarkable intricacy, laying bare craters that bear witness to cosmic collisions and towering mountains that punctuate the lunar terrain. In the pursuit of lunar observation, whether through the lenses of binoculars or the aperture of a telescope, the novice stargazer is bestowed with a mesmerizing encounter, an odyssey into the nuanced details etched upon the lunar visage. Allow me to expound upon the tapestry of lunar features, an invitation for beginners to embark on a cosmic odyssey of discovery.

Mare Imbrium (Sea of Showers), is one of the large dark areas on the Moon, forming a prominent feature. A vast lava plain with a diameter of about 1,200 kilometers. Look for the prominent mountain ranges surrounding it, such as the Montes Alpes. The Imbrium Basin formed from the collision of a proto-planet during the Late Heavy Bombardment. Basaltic lava later flooded the giant crater to form the flat volcanic plain seen today. The basin's age has been estimated using uranium–lead dating methods to approximately 3.9 billion years ago, and the diameter of the impactor has been estimated to be 250 ± 25 km.[2]

Embarking further on our lunar odyssey, we encounter yet another conspicuous lunar Mare, Mare Serenitatis, positioned to the celestial northeast of Mare Imbrium. Within its lunar contours, you will discover the captivating crater Posidonius, poised near its rim, a testament to the celestial forces that have sculpted the lunar landscape. Traverse further, and witness the Serpentine Ridge, an enigmatic feature etching its sinuous path across the Mare's center—a geological signature etched upon the lunar visage, beckoning the curious mind to unravel the cosmic narratives inscribed in its lunar tale.

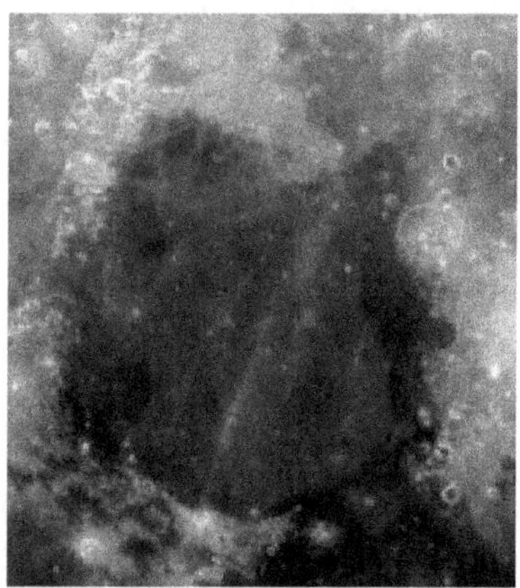

Nestled close to the northern periphery of Mare Imbrium, lies the celestial antiquity of Plato—a grand crater that has witnessed the eons unfold across the lunar landscape. This colossal lunar expanse, distinguished by its expansive, level floor, beckons the observer to peer into the profound mysteries etched upon its ancient surface.

Engage in a cosmic contemplation of Plato's dusky floor and the majestic rise of its central peaks—a testament to the celestial ballet of impacts and transformations. The allure of Plato lies in its juxtaposition—a relatively flat terrain amidst the rugged lunar expanse, an enigmatic anomaly that captures the discerning gaze and propels the stargazer into a cosmic reverie.

Tycho, a substantial and noteworthy crater ensconced within the lunar southern highlands is distinguishable by its luminous rays that gracefully traverse the lunar expanse, and stands as a celestial masterpiece capturing the imagination. This lunar marvel, relatively youthful in its cosmic existence, bears the hallmark of dynamic forces with a central peak that invites the discerning observer to partake in a meticulous exploration. Within Tycho's rim, a cosmic chronicle unfolds—a tale of impacts and upheavals etched upon the lunar canvas, awaiting the keen eye of the stargazer to unravel its celestial narrative.

Orion's Belt

Orion is a well-known constellation in the night sky, often called "The Hunter." It's visible during winter in the northern hemisphere and summer in the southern hemisphere. Orion is steeped in mythology across various cultures, and the constellation is marked by several bright stars, including Rigel and Betelgeuse, as well as the distinctive three-star belt known as Orion's Belt. One of the easiest ways to identify Orion is by finding Orion's Belt, which consists of three bright stars in a straight line. These stars are Mintaka, Alnitak, and Alnilam. Betelgeuse and Rigel, two bright stars with distinct colors, mark Orion's left shoulder and right foot, respectively. Bellatrix is another bright star on Orion's left shoulder, while Saiph is located at his right knee.

To locate Orion's Belt, look roughly midway between the eastern and western horizons. Once found, you can use the belt as a guide to explore other features within the constellation. Hanging from Orion's Belt is his Sword, including the famous Orion Nebula (M42), a stunning sight visible to the naked eye, binoculars, or a telescope. Additionally, there's Orion's Shield (Lambda Orionis) near the Belt, and if you're under dark skies, you might catch a glimpse of Barnard's Loop, a large emission nebula forming a loop around Orion's Belt.

In mythology, Orion was a great hunter in Greek lore, and various myths surround his adventures, including the pursuit of the Pleiades. Orion is a fantastic starting point for beginners due to its distinctive shape and bright stars. Exploring the wonders of the night sky through Orion opens the door to further celestial discoveries and makes stargazing an enjoyable experience for budding astronomers.

Sirius - the Dog Star

Sirius, also known as the Dog Star, is the brightest star visible from Earth and holds a prominent place in our night sky. Located in the constellation Canis Major, Sirius is approximately 8.6 light-years away from us, making it one of our closest stellar neighbors. Its luminosity is primarily attributed to its intrinsic brightness and proximity to our solar system.

To find Sirius, begin by identifying the prominent constellation Orion, recognizable by its distinctive three-star belt. If you draw a line through Orion's Belt from left to right and extend it southeastward, you'll encounter Sirius. During winter months in the northern hemisphere, Sirius becomes visible in the southeast soon after sunset, outshining all other stars.

- Sirius shines with a dazzling, pure white light, making it easily distinguishable from surrounding stars. Its brilliance is attributed to its close proximity and inherent brightness.

- As Sirius hangs low on the horizon, its light often appears to shimmer and change colors due to atmospheric effects. This twinkling phenomenon, known as stellar scintillation, adds to the star's allure.
- Sirius is a binary star system, consisting of a main-sequence star (Sirius A) and a faint white dwarf companion (Sirius B). However, Sirius B is challenging to observe without specialized equipment due to the stark brightness contrast with Sirius A.
- Sirius forms a notable part of the Winter Triangle, along with Betelgeuse in Orion and Procyon in Canis Minor. This celestial triangle becomes prominent during winter nights.
- In various cultures, Sirius has held cultural and mythological significance. In ancient Egypt, its annual rise heralded the flooding of the Nile, symbolizing renewal and prosperity.
- Even without binoculars or telescopes, Sirius is a striking object in the night sky. Its visibility and unmistakable brilliance make it an excellent target for beginner astronomers.
- Positioned in the constellation Canis Major, Sirius is often referred to as the Dog Star. The constellation itself outlines the shape of a large dog following the hunter Orion across the sky.

The Pleiades - A Celestial Cluster of Seven Sisters

One of the most famous star clusters is the Pleiades, also known as the Seven Sisters, located in the constellation of Taurus. This open cluster consists of several young, hot stars, and can be easily observed with the naked eye or binoculars. Pointing your telescope towards the Pleiades will reveal a stunning sight, where the stars appear like diamonds on a velvet black canvas.

To find the Pleiades, start by identifying the V-shaped cluster of stars that forms the head of Taurus, the Bull. Aldebaran, the red giant star, marks the eye of the bull. Move your gaze toward the opposite end of the V-shape, and you'll encounter the Pleiades. This star cluster is most visible during autumn and winter, and it stands out prominently against the surrounding celestial backdrop.

The Pleiades are an enchanting celestial object for beginner astronomers, offering a captivating blend of mythology, cultural significance, and visual appeal. Their accessibility and striking appearance make them a delightful target for those exploring the wonders of the night sky.

- The Pleiades consist of several hot, blue-white stars, and their collective glow creates a captivating spectacle. While traditionally known as the Seven Sisters, most people can discern six or seven stars with the naked eye.
- The Pleiades are often surrounded by faint nebulosity, known as reflection nebulae, created by the scattered light of the cluster's stars interacting with interstellar dust. This adds a delicate and ethereal

touch to the Pleiades' appearance.

- In Greek mythology, the Pleiades are associated with seven sisters who were pursued by the hunter Orion. To protect them, the gods transformed the sisters into stars, and they now reside in the night sky as the Pleiades.
- While the Pleiades appear close-knit from our vantage point, they are actually part of a loose cluster that extends over a larger volume of space. This grouping enhances the overall visual appeal of the star cluster.
- The Pleiades are visible to the naked eye, making them accessible and enjoyable for beginners. Their bright and distinct arrangement makes them a perfect starting point for those new to stargazing.
- The Pleiades hold cultural significance in various traditions around the world. They have been used as a navigational reference by different cultures and feature prominently in folklore and mythology.
- While the Pleiades are stunning to observe with the naked eye, they become even more spectacular when viewed through binoculars or a small telescope. The additional magnification reveals more individual stars and enhances the beauty of the cluster.

The Big Dipper

The Big Dipper, an iconic asterism within the constellation Ursa Major (the Great Bear), graces the night sky with its distinctive seven bright stars arranged in the shape of a celestial ladle. Easily recognized and visible from both hemispheres, the Big Dipper serves as a prominent guidepost for beginners exploring the wonders of the night sky.

The Big Dipper is an excellent starting point for novice astronomers, serving as a recognizable and easily located feature in the night sky. Its bright stars, distinctive shape, and navigational significance make it a celestial delight for those embarking on their stargazing journey.

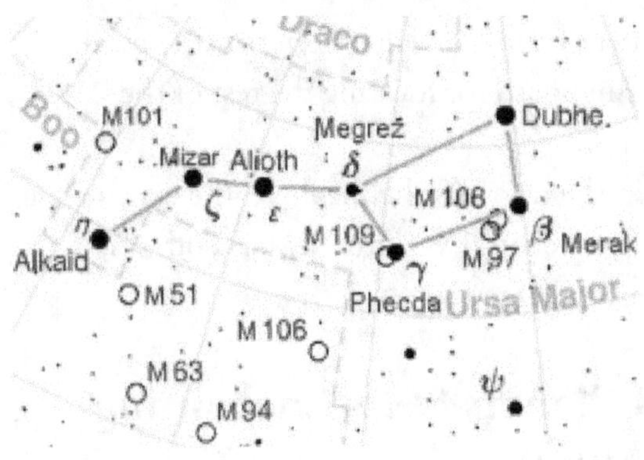

To find the Big Dipper, look toward the northern part of the sky during any season. In the northern hemisphere, it is often visible all year round, rotating counterclockwise around the North Star (Polaris). The Big Dipper is part of the larger Ursa Major constellation, and its handle points in the direction of the constellation's head. The two stars at the outer edge of the ladle's bowl, Merak and Dubhe, point towards Polaris, aiding in navigation.

- The Big Dipper's distinctive shape resembles a large ladle or saucepan. Its seven bright stars are easily identifiable and stand out against the surrounding stars.
- The stars of the Big Dipper are all relatively bright, making them easy to spot with the naked eye. The two stars at the outer edge of the bowl, Dubhe and Merak, are particularly prominent.
- The two stars at the end of the Big Dipper's bowl, Dubhe and Merak, point toward Polaris, the North Star. Polaris is located almost directly above Earth's North Pole and serves as a reference point for navigation.
- Alcor and Mizar, two stars in the handle of the Big Dipper, form a famous double star. While Alcor is fainter and requires keen eyes or binoculars to separate, Mizar itself is a binary star system.
- The orientation of the Big Dipper changes throughout the night, and its position relative to the horizon is a useful indicator of time. It has been a crucial navigational tool for various cultures throughout history.
- The Big Dipper is part of the larger Ursa Major constellation, which

resembles a bear. The Dipper's handle extends from the bear's tail, providing a convenient springboard for locating the rest of the constellation.

- While visible year-round in the northern hemisphere, the orientation and position of the Big Dipper change with the seasons, offering different perspectives at various times.

Jupiter: The King of Planets & Its Enchanting Moons

Jupiter, the largest planet in our solar system, reigns as the King of Planets with its imposing size and mesmerizing features. This gas giant is a captivating object for astronomers, both seasoned and beginner alike, and its system of moons adds to the allure of its celestial presence.

Finding Jupiter in the night sky is relatively straightforward, as it is one of the brightest objects visible to the naked eye. Typically, Jupiter can be observed in the eastern part of the sky after sunset, shining with a steady and radiant light. Its brightness often surpasses that of surrounding stars, making it stand out prominently. To pinpoint Jupiter, star maps, astronomy apps, or planetarium software can be valuable tools. Additionally, during opposition (when Jupiter is directly opposite the Sun in the sky), it is at its brightest and can be seen throughout the night.

Jupiter, with its immense size, captivating features, and entourage of moons, is an excellent target for beginners venturing into astronomy. Whether observed with the naked eye or through a telescope, this gas giant and its celestial companions offer a captivating glimpse into the vastness and beauty of our solar system.

- Jupiter's colossal size is its defining characteristic, with a diameter more than 11 times that of Earth. Its immense gravitational field contributes to its status as the solar system's unofficial guardian.
- Jupiter's dynamic atmosphere showcases prominent cloud bands of varying colors, including browns, reds, and whites. These bands are composed of different chemical compounds and atmospheric gases.
- A striking feature on Jupiter is the Great Red Spot, a massive storm that has raged for centuries. This enormous anticyclonic storm is larger than Earth and is easily visible through telescopes.
- Jupiter boasts a diverse family of 79 known moons, but the four largest, known as the Galilean moons, are particularly captivating. Io, Europa, Ganymede, and Callisto can be observed with binoculars or a small telescope.
- Io, the closest of the Galilean moons to Jupiter, is known for its volcanic activity and colorful surface markings. Its ever-changing appearance offers dynamic views for observers.
- Europa, with its icy surface, is believed to have a subsurface ocean. Its smooth appearance and potential for life make it a fascinating subject for study.
- Ganymede is the largest moon in the solar system, even larger than the planet Mercury. Its varied terrain includes dark regions, likely composed of rock and ice.
- Callisto, with its heavily cratered surface, is the most distant of the Galilean moons. Its ancient landscape provides insights into the history of the Jovian system.
- Observers may witness the dance of Jupiter's moons as they eclipse

one another or cast their shadows on the planet's surface. These events add an extra layer of intrigue to Jupiter's nightly display.
- Due to Earth's orbit, Jupiter's position in the night sky changes over the course of the year. Observing its changing location provides an opportunity to track its motion against the backdrop of stars.

Mars – The Red Planet

Mars, with its distinct color, surface features, and captivating moons, serves as an engaging celestial object for beginners venturing into astronomy. Whether observed with the naked eye or through a telescope, the Red Planet offers a glimpse into the mysteries of our neighboring planets and the cultural stories that have woven through human history.

Mars, often referred to as the Red Planet, commands attention in the night sky with its distinctive color and fascinating features. For beginners eager to explore the wonders of our solar system, Mars offers a captivating introduction to planetary observation.

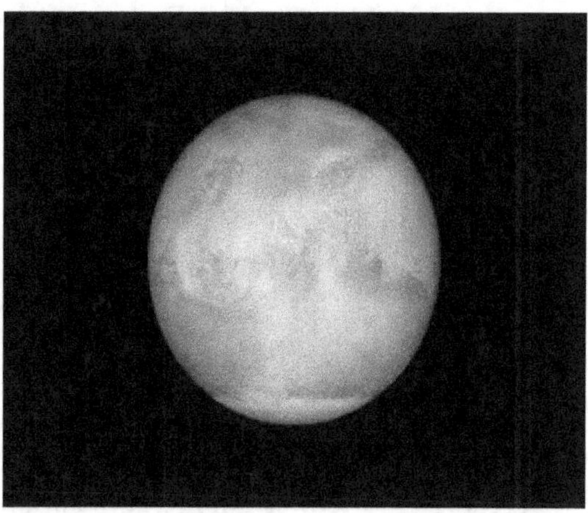

In ancient mythology, Mars was associated with the Roman god of war, identified with the Greek god Ares. This connection has influenced cultural references, naming conventions, and even the designation of Mars as the fourth planet from the Sun.

The reddish appearance of Mars is due to iron oxide, or rust, on its surface. This unique color makes it easily distinguishable from other celestial objects.

Through a telescope, observers can sometimes discern the polar ice caps on Mars. These ice caps undergo seasonal changes, expanding and contracting as the planet orbits the Sun.

Mars showcases various surface features, including dark regions, lighter plains, and expansive canyons. Notable among these is Valles Marineris, a vast canyon system that dwarfs Earth's Grand Canyon.

Olympus Mons, the largest volcano in the solar system, is a prominent feature on Mars. Its immense size and shield-like shape make it a captivating sight through telescopes.

Mars has two small moons, Phobos and Deimos. While challenging to observe with the naked eye, they can be seen with telescopes. Phobos, the larger moon, orbits closely to Mars, while Deimos orbits at a greater distance.

Tips for Observing Mars

- Use Binoculars or a Telescope: While Mars is visible to the naked eye, using binoculars or a telescope enhances the viewing experience, allowing for a closer look at surface details and the moons.
- Check Opposition Dates: Planetary oppositions, when Mars is opposite the Sun in the sky, provide optimal viewing opportunities. At opposition, Mars appears larger and brighter. Check online calendars for upcoming opposition dates.
- Observe During Opposition: During opposition, Mars rises as the Sun sets, making it visible throughout the night. (Mars is in the opposite direction from the sun). This is an ideal time for extended observations and detailed examination of the planet's features.

The Andromeda Galaxy

In Greek mythology, Andromeda was a princess chained to a rock as a sacrifice to appease a sea monster. She was later rescued by the hero Perseus, and they were immortalized as constellations. The Andromeda Galaxy's name is derived from this mythological figure.

Observing the Andromeda Galaxy is a memorable experience for beginners, offering a glimpse into the vast cosmic tapestry that extends beyond our own galaxy. As you explore its spiral beauty, consider the rich mythology that has woven these celestial wonders into the cultural fabric of human history.

The Andromeda Galaxy, also known as M31, stands as a celestial masterpiece and the closest spiral galaxy to our Milky Way. For beginners eager to explore the vastness of the universe, the Andromeda Galaxy provides a breathtaking introduction to deep-sky observation.

The Andromeda Galaxy boasts a stunning spiral structure, similar to our own Milky Way. Its graceful arms and central bulge make it a captivating sight through telescopes.

A prominent feature of M31 is its bright galactic core, where a dense concentration of stars resides. Observers can witness the glow emanating from this central region.

Andromeda is accompanied by two satellite galaxies, M32 and M110, which are visible in the same field of view through telescopes. These companions add to the overall cosmic tableau.

The Andromeda Galaxy is located approximately 2.5 million light-years away from Earth, making it the closest spiral galaxy to our own. Observing Andromeda provides a glimpse into the vast distances of our cosmic neighborhood.

Tips for Observing The Andromeda Galaxy

- To find the Andromeda Galaxy, look northeast during the fall and winter months in the Northern Hemisphere. Start by identifying the distinctive "W" shape of the constellation Cassiopeia, and then move to the left, following the stars to locate the constellation Andromeda.
- M31 appears as a faint, elongated smudge in the sky, visible to the naked eye under dark conditions.
- For optimal observation, choose locations away from city lights where the night sky is darker. This enhances the visibility of the Andromeda Galaxy.
- While visible to the naked eye, binoculars or a telescope reveal more details and the galaxy's intricate structure. Experiment with different magnifications to find the most appealing view.
- Astro-photographers can capture stunning images of the Andromeda Galaxy by taking long-exposure photographs. This technique reveals the galaxy's delicate spiral arms and intricate details.

The Summer Triangle

Observing the Summer Triangle is a delightful introduction to stargazing, providing a visually striking and easily recognizable formation in the night sky. As you trace the paths of Vega, Deneb, and Altair, consider the mythical tales that connect these celestial jewels, and let the beauty of the cosmos unfold before your eyes. While the Summer Triangle itself is not deeply rooted in mythology, its constituent stars have rich cultural significance. In Greek mythology, Vega is associated with the lyre of Orpheus, Deneb represents the tail of the celestial swan, and Altair is

linked to the eagle that carried Zeus's thunderbolts.

The Summer Triangle, an asterism formed by three bright stars from different constellations, graces the summer night sky, captivating the gaze of beginner astronomers with its prominent and easily identifiable shape. This stellar trio offers a delightful entry point for those exploring the wonders of the cosmos.

The Summer Triangle gracefully bisects the luminous band of the Milky Way, adding to its visual appeal. This intersection enhances the stargazing experience, especially under dark, clear skies.

Tips for Observing The Summer Triangle

- The Summer Triangle is most prominent during the summer months in the Northern Hemisphere, dominating the night sky. Look overhead in the evening to witness its splendor.
- To locate it, look for three prominent stars: Vega in the constellation Lyra (the Harp), Altair in Aquila (the Eagle), and Deneb in Cygnus (the Swan). These stars form a large, almost equilateral triangle that spans across the Milky Way.
- While visible to the naked eye, using binoculars or a telescope enhances the viewing experience. Telescopes can reveal additional

stars and details within the constellations.

- As the brightest star in the Summer Triangle, Vega stands out with its bluish-white brilliance. It is part of the constellation Lyra and is often associated with the mythical harp played by the Greek muse Erato.
- Deneb, situated in the constellation Cygnus, represents the tail of the celestial swan. This luminous, white supergiant star adds to the beauty of the Summer Triangle and is one of the most distant stars visible to the naked eye.
- Altair, the southernmost star in the trio, is found in the constellation Aquila. This yellow-white star is part of a prominent asterism known as the "Summer Diamond" and holds cultural significance in various traditions.

Lyra: The Celestial Harp in the Summer Sky

Lyra is a small but prominent constellation that graces the night sky during the summer months in the Northern Hemisphere. Named after the lyre, an ancient harp-like musical instrument, Lyra is rich in mythology and hosts several captivating celestial objects for the beginner stargazer.

In Greek mythology, Lyra is associated with the lyre played by the legendary musician Orpheus. The harp was said to produce music of such enchanting beauty that it could soothe wild beasts and even move inanimate objects. Vega, the brightest star in Lyra, is often linked to this mythical lyre and serves as a reminder of the celestial stories woven into the night sky.

Observing Lyra provides a delightful entry point for beginning stargazers, allowing them to explore bright stars, binary systems, and captivating nebulae. Whether appreciating the brilliance of Vega or seeking out the delicate beauty of the Ring Nebula, Lyra offers a celestial journey filled with myth and wonder in the summer night sky. Familiarize yourself with Lyra's distinctive shape, resembling a small harp or a parallelogram. Once recognized, Lyra becomes a familiar guidepost in the summer night sky. While Lyra's bright stars are visible to the naked eye, using binoculars or a telescope enhances the stargazing experience. This is particularly beneficial when observing objects like the Ring Nebula or resolving the Double-Double.

Some prominent celestial objects in Lyra visible to the beginning stargazer include:

Vega (Alpha Lyrae)

- As the brightest star in Lyra, Vega is a blue-white gem that anchors the constellation. Its brilliance makes it an easily identifiable point in the summer sky. Vega is also one of the stars that form the Summer Triangle.

Epsilon Lyrae (The Double-Double)

- Epsilon Lyrae, often known as the Double-Double, is a fascinating double-star system that can be resolved into two binary pairs through a telescope. This celestial gem showcases the beauty of stellar companionship.

The Ring Nebula (M57)

- Located between Beta and Gamma Lyrae, the Ring Nebula (M57) is a planetary nebula resembling a tiny, colorful smoke ring. Observing it through a telescope reveals its ethereal beauty, a result of a dying star shedding its outer layers.

Lyrid Meteor Shower Radiant

- The Lyrid meteor shower, which occurs annually in April, appears to originate from the vicinity of Lyra. While the shower is most active in April, meteors associated with the Lyrids can occasionally be observed throughout the year.

Aquila: Soaring High in the Summer Sky

Aquila, the celestial eagle, is a striking constellation that commands the summer night sky with its majestic presence. For the beginning stargazer, Aquila offers an opportunity to explore its prominent features and discover celestial objects within its boundaries.

In Greek mythology, Aquila is often associated with the eagle that carried Zeus's thunderbolts. The eagle played a crucial role in various myths, symbolizing strength and power. Additionally, Aquila is linked to the story of Ganymede, a Trojan prince who was abducted by Zeus in the form of an eagle and later became the cupbearer of the gods.

Aquila is best observed during summer evenings in the Northern Hemisphere, positioned high in the southeastern sky. Look for the distinctive pattern resembling an eagle with outstretched wings. The brightest star in Aquila is Altair, forming the southern tip of the Summer Triangle alongside Vega and Deneb. Learn to recognize the distinctive shape of Aquila, resembling a majestic eagle with widespread wings. This recognizable pattern makes it easier for beginners to locate and remember. Use Altair, the brightest star in Aquila and part of the Summer Triangle, as a guide to finding the constellation. Altair's brilliance aids in identifying the overall shape of the eagle. Optimal observation of Aquila and its celestial objects is achieved in dark sky locations away from city lights. This enhances visibility and allows for a clearer view of the night sky.

Observing Aquila offers a journey into both mythology and astronomy, combining the allure of ancient tales with the exploration of celestial wonders. Whether appreciating the wingspan of the celestial eagle or seeking out nebulae and star clusters, Aquila provides a captivating canvas for beginner stargazers to explore the mysteries of the summer night sky.

Some prominent celestial objects in Aquila visible to the beginning stargazer include:

Altair (Alpha Aquilae)

- Altair, a bluish-white star, is the brightest in Aquila and serves as a guiding beacon. Its location in the Summer Triangle makes it easily identifiable, and its rapid rotation adds a dynamic element to Aquila's celestial display.

Eagle Nebula (M16)

- While situated near Altair's vicinity, the famous Eagle Nebula (M16) is officially located in the neighboring constellation Serpens. However, it is easily visible from Aquila and renowned for its "Pillars of Creation," regions where new stars are born.

NGC 6751 (Glauert's Nebula)

- Glauert's Nebula, also known as NGC 6751, is a planetary nebula within Aquila. Though not as prominent as the Eagle Nebula, it is a captivating target for stargazers equipped with telescopes.

Aquila's Star Cluster (Messier 11)

- The Wild Duck Cluster, or Messier 11 (M11), is a vibrant open star cluster in Aquila. Its name is derived from the V-shaped pattern of stars resembling a flock of ducks. Binoculars or a telescope enhance the view, revealing the cluster's individual stars.

Cygnus: The Celestial Swan Gracefully Gliding Across the Night Sky

Cygnus, also known as the Northern Cross, is a captivating constellation that adorns the night sky during the summer months. Recognized for its distinctive shape resembling a flying swan, Cygnus offers an enchanting experience for beginning stargazers eager to explore the wonders of the cosmos.

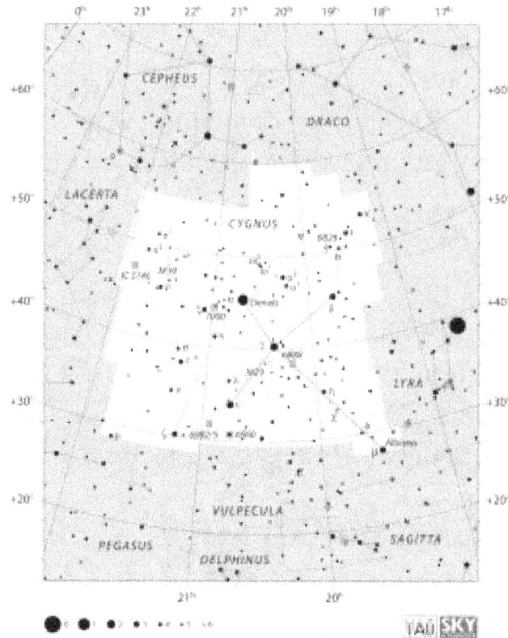

Observing Cygnus provides a delightful journey into both mythology and astronomy. From the recognizable shape of the celestial swan to the intricate details of nebulae and double stars, Cygnus invites beginning stargazers to explore the celestial wonders that grace the summer night sky.

In Greek mythology, Cygnus is often associated with several myths, including the tragic tale of the musician Orpheus. After his death, Orpheus was transformed into a swan and placed in the heavens as a constellation. Cygnus is also linked to the story of Zeus and Leda, where Zeus transformed into a swan to seduce Leda.

Cygnus is prominently visible in the Northern Hemisphere during summer evenings. Look for the prominent cross-shaped pattern formed by the constellation's bright stars. The star Deneb, part of the Summer Triangle along with Altair and Vega, marks the tail of the celestial swan. The neck stretches out toward the east, and the wings extend to the west, creating a striking and easily recognizable configuration. Identify the cross-shaped pattern formed by the bright stars in Cygnus. The Northern Cross is a distinctive and easily identifiable feature, helping beginners locate and remember the constellation. Utilize Deneb's brightness and its association with the Summer Triangle as a guide to finding Cygnus. Deneb's position at the tail of the swan facilitates the identification of the entire constellation.

Some prominent celestial objects in Cygnus visible to the beginning stargazer include:

Deneb (Alpha Cygni)

- Deneb, the brightest star in Cygnus, is a luminous blue-white supergiant. Its position in the Summer Triangle adds to its brilliance, making it a key anchor point for identifying the constellation.

Albireo (Beta Cygni)

- Albireo is a beautiful double star located at the head of the swan. Through binoculars or a telescope, Albireo reveals a striking color contrast with its golden and blue components, creating a captivating visual experience.

North America Nebula (NGC 7000)

- Shaped like the North American continent, the North America Nebula is a large emission nebula in Cygnus. While challenging to observe without a telescope, its presence adds to the celestial tapestry of the constellation.

Veil Nebula (NGC 6960)

- The Veil Nebula is a supernova remnant consisting of delicate filaments of gas. While extensive and faint, parts of the Veil Nebula can be observed with binoculars or a telescope, providing a glimpse into the aftermath of a stellar explosion.

Observing the Summer Triangle is a delightful introduction to stargazing, providing a visually striking and easily recognizable formation in the night sky. As you trace the paths of Vega, Deneb, and Altair, consider the mythical tales that connect these celestial jewels, and let the beauty of the cosmos unfold before your eyes.

Meteor Showers and Celestial Fireworks

In the cosmic fireworks display, meteor showers steal the spotlight. Let's explore the ephemeral beauty of shooting stars and unlock the secrets of celestial phenomena.

In the cosmic tapestry of our universe, meteor showers emerge as celestial choreographies, captivating the gaze of stargazers and weaving themselves into the cultural heritage of diverse civilizations. Born from the remnants of celestial wanderers—comets and shattered asteroids—these luminous phenomena engage in a cosmic ballet with Earth as it traverses celestial pathways. Named after constellations like Perseus and Gemini, these meteor showers offer a celestial feast, enhancing our connection to the wonders of the cosmos. From the ethereal beauty of the Perseids, associated with the tears of Saint Lawrence, to the gemstone-like Geminids linked to asteroid 3200 Phaethon, each shower tells a unique story in the cosmic theater.

The cosmic ballet involves the interplay of Earth, cometary debris, and the dance of celestial bodies, deepening our connection to the vastness of space. Advances in astronomy empower us to predict meteor shower activity, turning casual stargazing into a meaningful exchange with the universe. Fireballs and bolides add an unpredictable yet awe-inspiring dimension to this cosmic symphony, showcasing the dynamic nature of celestial encounters. Meteor showers, with their cultural significance, become cosmic messengers, reminding us of the interconnectedness of the universe and inviting us to embrace the ephemeral beauty of the celestial spectacle. As shooting stars trace luminous arcs across the night sky, we are urged to join in the timeless ode to the wonders of the universe—a grand stage where cosmic wonders unfold, leaving an indelible mark on the human imagination.

Celestial Debris: Origins of Meteor Showers

The mesmerizing celestial ballet of meteor showers is a harmonious interplay between Earth, cometary debris, and the dance of celestial bodies. At its core, this cosmic phenomenon involves the encounter between our planet and debris left behind by comets or asteroids. As Earth traverses its orbit around the Sun, it moves through regions of space containing these celestial remnants. The collision between our atmosphere and these tiny particles results in the luminous streaks we observe as meteors.

Meteor showers showcase the dynamic nature of our cosmic environment. Cometary debris, composed of dust and rocks shed by passing comets, becomes a transient spectacle when it intersects with Earth's atmosphere. The celestial bodies involved, such as comets and asteroids, add an additional layer of complexity to this cosmic dance. Each meteor shower is a unique performance, with its own set of cosmic actors contributing to the dazzling display overhead.

Understanding the dynamics of meteor showers goes beyond the scientific aspects; it fosters a profound connection to the cosmos. Witnessing a meteor shower becomes more than a visual treat—it transforms into a shared experience with the universe. The awareness that these luminous streaks are the result of interactions between celestial bodies and our home planet enhances our cosmic perspective. It instills a sense of interconnectedness with the vastness of space and time, making the night sky a living canvas where celestial phenomena unfold.

The ability to predict meteor shower activity is a testament to the advancements in astronomy. Scientists utilize sophisticated models and observational data to anticipate when and where meteor showers will peak. These predictions take into account factors such as the orbit of comets, Earth's trajectory, and the density of debris in space. Such accuracy allows astronomers and enthusiasts alike to plan their observations effectively, ensuring they are in the right place at the right time to witness the celestial show.

As we delve into the dynamics of meteor showers, we become active participants in the cosmic conversation. Armed with knowledge about the celestial ballet, observers can appreciate the transient beauty of meteors with a deeper understanding of their origin. This engagement transforms a casual stargazing session into a meaningful exchange with the cosmos. Whether you're a seasoned astronomer or a beginner, grasping the intricacies of meteor shower dynamics enriches the experience, turning a night of observation into a cosmic journey through the wonders of our celestial home.

Meteor showers, with their origins traced to the remnants of comets and asteroids, offer a unique window into the cosmic past. The particles creating these luminous trails have traveled through the vastness of space, preserving a record of the solar system's history. Observing a meteor shower is, in essence, witnessing a celestial time capsule as these cosmic messengers ignite in our atmosphere. This realization adds a layer of fascination to meteor showers, inviting us to contemplate the cosmic narratives written in the night sky.

Earth's Celestial Light Show: Meteor Shower Basics

The journey begins with tiny cosmic travelers known as meteoroids. As they enter Earth's atmosphere, they become meteors, producing streaks of light. If a meteor survives its fiery passage and reaches Earth's surface, it earns the title of meteorite. Meteor showers reach their peak on specific nights, marked by radiant points in the sky. Observing meteors from dark-sky locations enhances the viewing experience.

Observing both the Perseid and Geminid meteor showers is an accessible and enchanting way for beginner astronomers to experience the beauty of the night sky. Whether you choose August or December, these meteor showers provide a captivating introduction to the wonders of our cosmic neighborhood.

Perseid Meteor Shower: Celestial Fireworks in August

The Perseids, associated with Perseus, have historical and cultural significance. Legend attributes their origin to the tears shed by Saint Lawrence, a Christian martyr. The celestial dance of Perseus becomes a poignant narrative that transcends scientific understanding. The Perseids' peak in August offers a celestial feast for stargazers.

The Perseid meteor shower is one of the most anticipated annual celestial events, captivating beginner astronomers with its dazzling display of shooting stars. These meteors, originating from the debris of Comet Swift-Tuttle, create a spectacular show in the night sky.

The Perseid meteor shower is named after the constellation Perseus, as the meteors appear to radiate from this point in the sky. In Greek mythology, Perseus was a hero known for slaying the Gorgon Medusa. The shower's association with Perseus adds a mythical layer to the celestial spectacle.

The Perseid meteor shower occurs annually in August, typically peaking around the nights of August 11-13. During this time, Earth passes through the debris trail left by Comet Swift-Tuttle, resulting in an increased number of meteors visible to observers.

Look towards the northeastern part of the sky to locate the radiant point in the constellation Perseus. While the meteors will streak across the entire sky, tracing their paths back to Perseus can enhance the experience.

The best time to observe the Perseids is during the pre-dawn hours when the sky is darkest. However, meteors can be visible as early as a few hours after sunset.

Tips for Observing the Perseids

- For a beginning astronomer eager to witness the celestial spectacle of the Perseid meteor shower in the northern hemisphere, timing and location are crucial. The Perseids typically peak in mid-August, making this period the optimal time for observation. During this time, the Earth passes through the densest debris fields left by the comet Swift-Tuttle, resulting in an increased meteor shower activity. It's essential to plan your observation night during the peak, which usually occurs around August 11-13.
- Dark-Sky Locations for Clear Views: Selecting an ideal location is equally important for an unforgettable Perseid meteor shower experience. Light pollution from urban areas can significantly hinder visibility, so it's recommended to choose a dark-sky location away from city lights. National parks, rural areas, or dedicated stargazing sites offer the best conditions for clear views of the night sky. Finding an elevated vantage point enhances your visibility, allowing you to witness more meteors streaking across the heavens.
- Observing Tools and Techniques: While the Perseids are visible to the naked eye, enhancing your experience with simple observing tools can make the event even more enjoyable for a beginner. Bring a reclining chair or a blanket to lie down comfortably, as meteor watching involves looking up for extended periods. Binoculars are useful for observing details like persistent trains—glowing ionized gas left by a meteor. However, for meteor showers, it's often best to rely on the unaided eye to catch the full expanse of the night sky.
- Patience and Adaptability: Observing meteor showers requires patience, as the meteors can appear sporadically. Plan to spend at least

a couple of hours under the night sky to increase your chances of catching more meteors. Additionally, be prepared for changing weather conditions. Check the weather forecast and be adaptable—clear skies contribute to optimal viewing conditions. If clouds obstruct your view, consider waiting for breaks or exploring nearby areas with clearer skies.

- Celestial Orientation and Radiant Point: Understanding the celestial choreography of the Perseids adds an extra layer of appreciation. The meteors of the Perseid shower appear to radiate from a point in the constellation Perseus, known as the radiant point. While you don't need to focus on a specific direction to see the meteors, locating Perseus in the night sky can be a delightful aspect of your observation. Use astronomy apps or star charts to familiarize yourself with the constellation's position, allowing you to identify the radiant point and appreciate the cosmic dance of the Perseids.

Geminid Meteor Shower: December's Celestial Gems

The Geminids, originating from Gemini, emerge as one of the most prolific meteor showers. Unlike many associated with comets, the Geminids are linked to the asteroid 3200 Phaethon. Their diverse colors and slow-moving trails distinguish them as celestial gems. The Geminids' radiant beauty and record-breaking displays captivate astronomers.

The Geminid meteor shower, known for its abundance of bright and colorful meteors, is a stunning celestial event that graces the December night sky.

The Geminid meteor shower gets its name from the constellation Gemini, as the meteors appear to radiate from this point in the sky. In Greek mythology, Gemini represents the twins Castor and Pollux, the offspring of Leda and Zeus. The celestial connection adds a mythic touch to the experience of watching the Geminid meteor shower.

The Geminid meteor shower typically reaches its peak around December 13-14 each year. During this time, Earth passes through the debris trail left by asteroid 3200 Phaethon, resulting in a shower of meteors visible across the sky.

Tips for Observing the Geminids

- Choosing the Optimal Time for Geminids Observation: For a novice astronomer eager to delve into the captivating world of meteor showers, the Geminids provide an excellent opportunity, typically reaching their peak in mid-December. Scheduling your observation during the peak nights, which usually occur around December 13-14, ensures a higher likelihood of witnessing an abundance of meteors. The Geminids are known for their reliability and intensity, making them a spectacular celestial event for those new to stargazing.

- Understanding the Geminids' Radiant Point: Enhance your Geminids meteor shower experience by understanding the celestial choreography associated with it. The meteors of the Geminids shower appear to radiate from a specific point in the sky, known as the radiant point. This radiant is situated in the Gemini constellation, hence the name of the meteor shower. While you don't need to focus on a specific direction to witness meteors, locating Gemini in the night sky can add an extra layer of appreciation to your observation. Utilize astronomy apps or star charts to familiarize yourself with Gemini's position, allowing you to identify the radiant point and immerse yourself in the cosmic dance of the Geminids.

Other Notable Meteor Showers

The Leonids, associated with the constellation Leo, deliver a celestial performance with a rich history, known for periodic meteor storms. The Quadrantids present a unique mystery regarding their parent body. Their brief but intense displays mark them as a celestial enigma worthy of exploration.

- Timing the Leonid Meteor Shower Observation: For budding astronomers eager to witness a celestial spectacle, the Leonid meteor shower offers a captivating experience typically around mid-November. The peak nights, often occurring between November 17-18, mark the optimal time to observe the Leonids. This meteor shower is renowned for occasional meteor storms, during which the rate of meteors increases significantly, adding an element of unpredictability to the observation.

- Favorable Viewing Conditions: Choosing the right time and conditions for observing the Leonid meteor shower is crucial for an enjoyable experience. The best viewing hours typically occur after midnight when the radiant point, the spot in the sky from which the meteors appear to originate, gains more altitude. Selecting a location away from light pollution, such as rural areas or designated stargazing sites, enhances visibility. Clear skies are essential, so monitoring the weather forecast in advance allows for better preparation and increased chances of optimal viewing conditions.

- Essential Tools for Observation: Observing the Leonids doesn't require specialized equipment, making it accessible for beginners. The naked eye is sufficient to witness the majority of meteors streaking across the night sky. However, bringing a reclining chair or a blanket for comfort is advisable, as meteor watching involves looking up for extended periods. Patience is key, as meteors can appear sporadically, and dedicating a substantial amount of time to your observation session increases the likelihood of witnessing more celestial phenomena.

- Understanding the Radiant Point: Enhancing your experience of the Leonid meteor shower involves understanding the concept of the radiant point. The meteors of the Leonids seem to radiate from a specific location in the sky, located within the Leo constellation, hence the name of the meteor shower. While you don't need to focus on a specific direction to see meteors, familiarizing yourself with Leo's position using astronomy apps or star charts adds an extra layer of appreciation to your observation. Identifying the radiant point allows you to immerse yourself in the cosmic dance of the Leonids.

- Adapting to Meteor Storm Possibilities: The Leonids are known for occasionally producing meteor storms, characterized by a significant increase in meteor activity. While meteor storms are unpredictable, being aware of this possibility adds an exciting dimension to your observation. If a storm is anticipated, it might be worthwhile to plan for an extended observation period during the peak nights, increasing the chances of witnessing an enhanced display. Stay flexible, as meteor storms are infrequent but can create unforgettable astronomical experiences for those fortunate enough to witness them.

Fireballs and Bolides: Spectacular Meteoric Phenomena

Occasionally, a meteor achieves the status of a fireball—an exceptionally bright and luminous meteor. Bolides, meteors that explode in a burst of light, highlight the dynamic nature of celestial encounters. Meteoroid explosions, often accompanied by sonic booms, add an element of drama to the cosmic theater.

In the mesmerizing tapestry of celestial events, a meteor occasionally ascends to extraordinary brilliance, earning the prestigious title of a fireball. A fireball is not just any meteor; it is an exceptionally bright and luminous celestial traveler that captivates observers with its intense glow. These cosmic wonders are a testament to the dynamic and unpredictable nature of celestial encounters, turning an ordinary meteor into an extraordinary spectacle.

Bolides, another captivating facet of meteoric phenomena, take the celestial drama to new heights. Unlike typical meteors that gracefully streak across the sky, bolides explode in a burst of brilliant light, leaving a luminous trail in their wake. These explosive displays showcase the incredible energy released during the meteoric encounter with Earth's atmosphere. The sudden and intense illumination of a bolide adds an element of theatrical grandeur to the cosmic theater, transforming the night sky into a canvas of celestial fireworks.

Meteoroid explosions, accompanied by sonic booms, bring an additional layer of drama to the cosmic stage. When a meteoroid, the precursor to a meteor, undergoes a dramatic explosion, the release of energy creates a shockwave that generates a sonic boom. This phenomenon intensifies the sensory experience of observing celestial events, adding an audible component to the visual spectacle. The combination of a bright flash, a trail of light, and the sonic resonance of a meteoroid explosion creates a multisensory cosmic encounter that lingers in the memory of those fortunate enough to witness it.

For the beginning astronomer, grasping the nature of fireballs and bolides enhances the appreciation of meteoric phenomena. These extraordinary events are a reminder of the dynamic and ever-changing nature of the cosmos. While most meteors quietly traverse the night sky, fireballs and bolides inject an element of surprise and awe, inviting observers to anticipate the unexpected in their celestial observations.

Observing fireballs and bolides reinforces the idea that the night sky is full of surprises. The unpredictable nature of these events adds an element of excitement to stargazing, encouraging astronomers to keep a watchful eye on the heavens. Whether you're a novice or seasoned stargazer, the allure of fireballs and bolides lies in their ability to transform a routine night of observation into an extraordinary celestial experience. These luminous displays serve as a vivid reminder that, in the cosmic theater, each meteor has the potential to become a star of its own spectacular show.

Meteor Showers in Cultural Heritage

Meteor showers, with their ethereal beauty, hold cultural significance across diverse civilizations. From ancient myths and folklore to artistic interpretations, meteor showers have left an indelible mark on human culture. Artists and writers throughout history have drawn inspiration from the celestial phenomena of meteor showers.

In the annals of cultural heritage, meteor showers find a place in ancient myths and folklore. These celestial events were often interpreted as omens, portending significant events or foretelling the fates of individuals and societies. The tears of the stars, as manifested in meteor showers, became part of intricate narratives that shaped the belief systems of ancient cultures. Understanding the cultural symbolism embedded in these celestial displays provides a unique lens through which we can explore the rich tapestry of human mythology.

Meteor showers have served as a muse for artists throughout history, inspiring creativity across various forms of artistic expression. From paintings to literature, the luminous trails of meteors have found their way into the works of renowned artists and writers. The visual spectacle of meteor showers, with their fleeting yet intense brightness, has been captured on canvas and in the written word, adding a celestial dimension to the artistic legacy of humanity.

The representation of meteor showers in art and literature stands as a testament to the enduring impact of these cosmic events on the human imagination. Artists have sought to convey the beauty and mystique of meteor showers, while writers have incorporated them into stories that explore the interconnectedness of the cosmos and the human experience. Meteor showers, as depicted in various art forms, become not only celestial phenomena but also metaphors for the transient and sublime aspects of life.

Beyond artistic representation, meteor showers have played a role in shaping cultural rituals and traditions. Across cultures, stargazers have engaged in practices during meteor showers, such as making wishes upon shooting stars or organizing meteor shower parties. These rituals reflect the enduring sense of wonder and magic associated with celestial events, showcasing how meteor showers have become a source of inspiration, contemplation, and communal celebration throughout the ages.

In essence, meteor showers serve as cosmic messengers, delivering a profound message from the universe to humanity. Their transient beauty and cultural significance highlight the interconnectedness of the cosmos and the cosmic forces that shape our celestial home. For the beginning astronomer, delving into the cultural heritage of meteor showers provides not only a richer understanding of their astronomical nature but also a deeper connection to the timeless and universal aspects of human culture.

Meteor Showers: Messages from the Cosmos

The transient beauty of meteor showers carries a profound message from the cosmos. These celestial displays, ephemeral yet impactful, remind us of the interconnectedness of the universe and the cosmic forces that shape our celestial home. Stargazers across cultures have engaged in rituals and traditions during meteor showers, reflecting the enduring sense of wonder and magic associated with celestial events.

Embracing the Celestial Spectacle

As we conclude our exploration of meteor showers, one undeniable truth emerges—the cosmos is a grand stage where ephemeral wonders unfold. Meteor showers invite us to embrace the celestial spectacle that has captivated humanity for centuries. Join me in the cosmic theater, where shooting stars trace luminous arcs across the night sky, and the ephemeral beauty of celestial phenomena becomes a timeless ode to the wonders of the universe.

Chapter 5: Moonlit Marvels - Lunar Phases and Features

In the celestial drama, the moon takes center stage. Let's explore the ever-changing phases and intricate features of Earth's cosmic companion.

Artists throughout history have sought to capture the moon's allure in their creations. From paintings and literature to music and dance, the moon becomes a muse for human expression. The artistic representations of lunar phases and features become a testament to the enduring fascination with Earth's cosmic partner.

In the cosmic drama, the moon takes center stage, inspiring human expression through art and literature. Lunar phases, from the waxing crescent to the waning crescent, create a rhythmic ballet as sunlight and shadows dance on the lunar surface. Observing the moon's eight distinct phases, one can trace the interplay of sunlight, unveiling the moon's monthly journey. Armed with a telescope, stargazers can delve into the moon's topography, observing craters, mountains, and maria, enhancing their cosmic exploration.

Exploring lunar features reveals a captivating lunar landscape marked by craters, maria, mountains, and rilles. Notable craters like Copernicus and Tycho narrate the moon's cosmic biography through celestial collisions. Lunar maria, once believed to be seas, offer a geological mosaic, contrasting dark lava plains with brighter highlands. Cultural perspectives add depth to our cosmic connection with Earth's celestial companion, enriching the lunar exploration experience.

Essential tips for lunar observation include choosing clear nights during waxing or waning gibbous phases, using binoculars or a telescope for enhanced detail, and familiarizing oneself with lunar geography through maps or apps. Lunar observation journals and NASA's resources further aid enthusiasts. The moon's gravitational influence extends to Earth's tides, shaping coastlines in a celestial dance. Human and robotic lunar exploration, coupled with lunar anomalies and Earthrise moments, contribute to our evolving understanding of the moon's mysteries.

The Apollo missions, robotic explorations, and lunar anomalies enhance our understanding of the moon's geological and cosmic history. Earthrise images from the lunar surface shift cosmic perspectives, symbolizing interconnectedness and environmental awareness. Future lunar exploration, exemplified by the Artemis program and the Lunar Gateway, holds the promise of scientific endeavors and international collaboration, inviting humanity to unlock the remaining secrets of Earth's cosmic companion. In this exploration of Moonlit Marvels, the moon's ever-changing phases and intricate features reveal cosmic secrets, fostering a profound connection between humanity and its celestial neighbor.

The Phases of the Moon

As the moon orbits Earth, it undergoes a mesmerizing transformation known as lunar phases. From the waxing crescent to the full moon and the waning crescent, the interplay of sunlight and shadows on the lunar surface creates a celestial ballet that unfolds in a rhythmic cycle. One of the most fascinating aspects of astronomy is observing the phases of the moon. The moon, our nearest celestial neighbor, goes through a series of distinct phases, each offering a unique and awe-inspiring view to the stargazers. We will now explore the different phases of the moon, their significance, and how they can be observed.

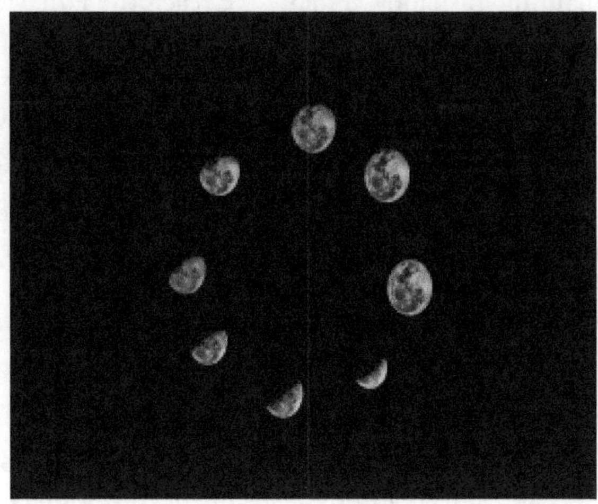

The moon has eight distinct phases: New Moon, Waxing Crescent, First Quarter, Waxing Gibbous, Full Moon, Waning Gibbous, Last Quarter, and Waning Crescent. These phases occur due to the relative positions of the sun, moon, and Earth. As the moon orbits the Earth, the sunlight falling on it changes, resulting in the various phases.

During the New Moon phase, the moon is completely hidden from view. It gradually emerges as a thin crescent during the Waxing Crescent phase. As it continues to wax, more of its surface becomes visible, leading to the First Quarter phase. The Waxing Gibbous phase follows, with only a small portion of the moon remaining in shadow. Finally, we reach the Full Moon, where the moon appears as a complete disc.

Understanding the dynamics of lunar phases involves tracing the path of sunlight across the moon's surface. The moon's orientation relative to the sun determines the amount of its illuminated side visible from Earth, offering stargazers a visual narrative of the moon's monthly journey through its phases. During the New Moon phase, the moon is completely hidden from view. It gradually emerges as a thin crescent during the Waxing Crescent phase. As it continues to wax, more of its surface becomes visible, leading to the First Quarter phase. The Waxing Gibbous phase follows, with only a small portion of the moon remaining in shadow. Finally, we reach the Full Moon, where the moon appears as a complete disc.

After the Full Moon, the moon starts to wane. The Waning Gibbous phase shows a decreasing illuminated area, followed by the Last Quarter phase where half of the moon is visible in the sky. The Waning Crescent phase marks the moon's return to invisibility, signaling the start of a new lunar cycle.

Observing these phases can be a thrilling experience, especially for beginners. Armed with a basic telescope, you can witness the moon's transformation night after night. Each phase offers a unique opportunity to observe the moon's craters, mountains, and maria (dark, flat areas). By studying the changing phases, you can develop a deeper understanding of the moon's topography and its relationship with the Earth and the sun.

For those new to stargazing, it's important to note that the moon's phases follow a predictable pattern and can be easily tracked using astronomy apps, calendars, or websites. This knowledge will help you plan your stargazing sessions, ensuring you don't miss any exciting lunar events.

NASA has developed and provided the "Daily Moon Guide[1]", which is a fully interactive guide to observing the Moon. You can also download and record your own personal Moon Observation Journal here[2] and printing it out for your own use.

And, if it's cloudy, rainy or just plain too cold to go outside to view the Moon, you can take a virtual tour of the Moon by clicking here[3] to visit NASA's Moon Trek! Moon Trek is an interactive Moon map made using NASA data from our lunar spacecraft. Fly anywhere you'd like on the Moon, calculate the distance or the elevation of a mountain to plan your lunar hike, or layer attributes of the lunar surface and temperature.

The moon's phases are a captivating aspect of astronomy that offers endless wonder and exploration. Whether you are a new amateur astronomer, a

---

1. https://moon.nasa.gov/moon-observation/daily-moon-guide/

2. https://moon.nasa.gov/resources/12/moon-observation-journal/

3. https://trek.nasa.gov/moon/

teacher, or a curious kid, observing and understanding the phases of the moon will enhance your stargazing experience and deepen your appreciation for the celestial wonders above. So grab your telescope, find a cozy spot under the night sky, and embark on a journey to unravel the mysteries of the moon's phases.

Lunar Features to Observe

As you embark on your journey into the captivating world of astronomy, one of the first celestial objects you will encounter is our very own Moon. The Moon has been a source of wonder and inspiration for humans for thousands of years, and observing its distinctive features can be an exciting and rewarding experience for stargazers of all ages. In this subchapter, we will explore some of the lunar features that you can look for while observing the Moon.

The moon's surface, marked by a vast cosmic history, is adorned with craters of various sizes. These impact scars tell the story of celestial collisions that shaped the lunar landscape. From the majestic Copernicus to the intricate Tycho, each crater unveils a chapter in the moon's cosmic biography. These circular depressions are formed by the impact of meteoroids, asteroids, and comets over billions of years. Craters come in various sizes, and some of the most famous ones can be easily observed using even a small telescope. Take note of their central peaks and the rays of ejected material that extend from them.

Tycho, a substantial and noteworthy crater ensconced within the lunar southern highlands is distinguishable by its luminous rays that gracefully traverse the lunar expanse, and stands as a celestial masterpiece capturing the imagination. This lunar marvel, relatively youthful in its cosmic existence, bears the hallmark of dynamic forces with a central peak that invites the discerning observer to partake in a meticulous exploration. Within Tycho's rim, a cosmic chronicle unfolds—a tale of impacts and upheavals etched upon the lunar canvas, awaiting the keen eye of the stargazer to unravel its celestial narrative.

Plato Crater, nestled close to the northern periphery of Mare Imbrium, lies the celestial antiquity of Plato—a grand crater that has witnessed the eons unfold across the lunar landscape. This colossal lunar expanse, distinguished by its expansive, level floor, beckons the observer to peer into the profound mysteries etched upon its ancient surface. Engage in a cosmic contemplation of Plato's dusky floor and the majestic rise of its central peaks—a testament to the celestial ballet of impacts and transformations. The allure of Plato lies in its juxtaposition—a relatively flat terrain amidst the rugged lunar expanse, an enigmatic anomaly that captures the discerning gaze and propels the stargazer into a cosmic reverie.

Another intriguing lunar feature is the maria, dark and flat areas on the Moon's surface. These regions were once believed to be seas, hence the name "maria," but they are actually vast solidified lava plains. The most famous maria, such as the Sea of Tranquility (where Apollo 11 landed) and the Sea of Serenity, can be seen with the naked eye or binoculars. Observe the contrast between these dark areas and the brighter highlands surrounding them.

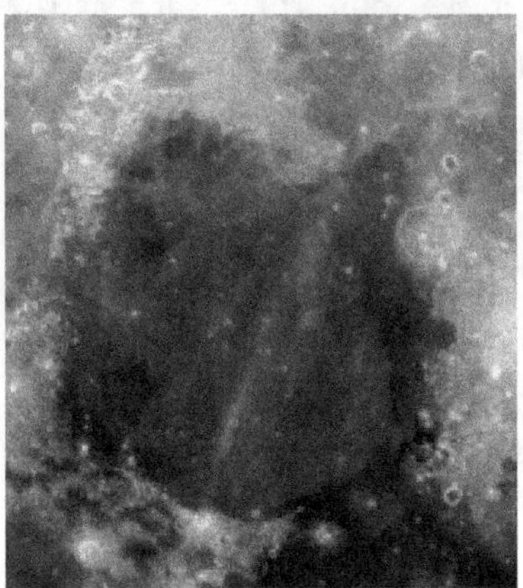

The Moon also offers a stunning display of mountains and valleys. The lunar mountains, including the towering range known as the Apennines, can be observed along the Moon's terminator, the line separating the illuminated and dark portions of the Moon. Valleys, like the dramatic Alpine Valley, provide a glimpse into the Moon's geological history and can be captivating to explore visually.

Additionally, be sure to keep an eye out for the lunar features known as rilles. These are long, narrow channels that resemble dried-up riverbeds and can be found in various locations on the Moon's surface. Some rilles, such as the Hadley Rille near the Apollo 15 landing site, offer a fascinating glimpse into the Moon's volcanic past.

Exploring these lunar features will not only deepen your understanding of our celestial neighbor but also enhance your stargazing experience. Remember to use a good lunar map or smartphone app to help you identify and locate these features. By observing the Moon regularly, you will witness its changing phases and gain a deeper appreciation for the wonders of our universe. Happy stargazing!

Lunar Eclipses

In the grand tapestry of the cosmos, there exists a celestial ballet, a gravitational dance that unfolds with precision and grace—the phenomenon of lunar eclipses. As we cast our eyes skyward, we witness the intricate interplay of Earth, Moon, and Sun, a cosmic choreography governed by the laws of celestial mechanics. This section embarks on a journey into the heart of this celestial ballet, exploring the mesmerizing world of lunar eclipses and unraveling the cosmic mysteries that unfold during these cosmic events.

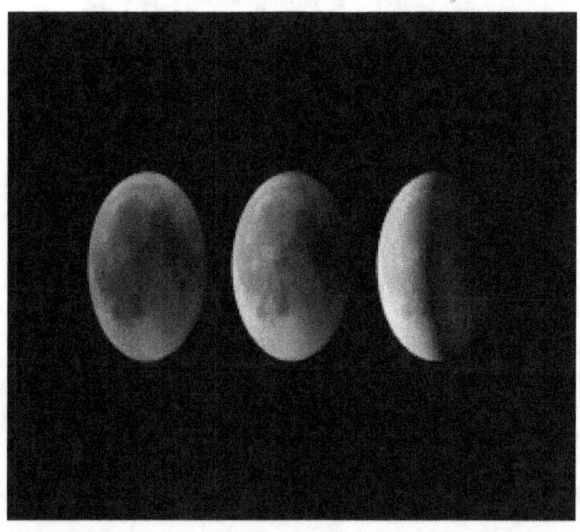

STARRY NIGHTS: A BEGINNER'S JOURNEY INTO ASTRONOMY

A lunar eclipse is a captivating celestial event that occurs when the Earth passes directly between the Sun and the Moon, creating a mesmerizing alignment known as syzygy. As the Earth casts its shadow onto the lunar surface, the Moon can take on various stunning hues, ranging from a subtle darkening to a vibrant copper or reddish tint. This captivating transformation is a result of the Earth's atmosphere filtering and bending sunlight, allowing only certain wavelengths to reach and illuminate the Moon. There are two main types of lunar eclipses: penumbral, where the Moon passes through the Earth's penumbral shadow, and umbral, where it moves through the darker, central part of the Earth's shadow. The latter often results in a more dramatic and visually striking eclipse, with the Moon turning a rich red color during totality. Lunar eclipses are observable from any location on the nighttime side of the Earth and offer a unique opportunity for stargazers and astronomers to witness the dance of celestial bodies in our cosmic neighborhood.

Our first act delves into the gravitational choreography that orchestrates lunar eclipses. Here, we illuminate the celestial mechanics behind these captivating phenomena, unraveling the intricate forces that bind our planet to its celestial companion. Understanding Earth's shadow and its dynamic interaction with the Moon unveils the cosmic mechanics that transform a seemingly ordinary lunar night into a celestial spectacle.

As we venture deeper into the cosmic narrative, we encounter the rich tapestry of folklore and cultural significance woven around lunar eclipses. Across diverse civilizations, myths and stories emerge, each adding a unique layer to the collective human understanding of these celestial occurrences. Lunar eclipses, throughout history, have not merely been astronomical events; they have been omens, symbols, and sources of inspiration that have left an indelible mark on the cultural fabric of humanity.

In the subsequent sections, we will delve into the narratives of diverse civilizations, exploring the myths and stories that have emerged from different corners of the world. From ancient tales passed down through generations to the contemporary interpretations that enrich our cultural tapestry, we will traverse the landscape of human imagination, discovering the myriad ways lunar eclipses have been perceived, revered, and integrated into the collective consciousness.

So, join me on this cosmic odyssey—a journey through the shadows, a dance of celestial bodies, and a voyage into the profound cultural meanings attached to lunar eclipses. As we embark on this exploration, we will unravel the secrets of the universe and contemplate the timeless beauty that unites the Earth, Moon, and Sun in a celestial embrace.

Celestial Ballet: The Gravitational Choreography of Lunar Eclipses

In the cosmic ballet of our celestial neighborhood, few events capture the imagination like a lunar eclipse. This mesmerizing dance unfolds as Earth, the Moon, and the Sun align in a celestial configuration that transcends mere astronomical mechanics. Understanding the intricate celestial choreography requires delving into the realms of celestial mechanics, where the cosmic forces governing our solar system reveal their captivating influence.

At the heart of a lunar eclipse lies the gravitational dance between Earth and its natural satellite. The Moon, orbiting our planet, occasionally finds itself immersed in Earth's shadow. This alignment, orchestrated by the delicate interplay of gravitational forces, leads to the captivating phenomenon we call a lunar eclipse. The intricate mechanics of this celestial ballet involve the precise positioning of Earth, Moon, and Sun, as they engage in a harmonious arrangement that briefly veils the lunar surface in Earth's shadow.

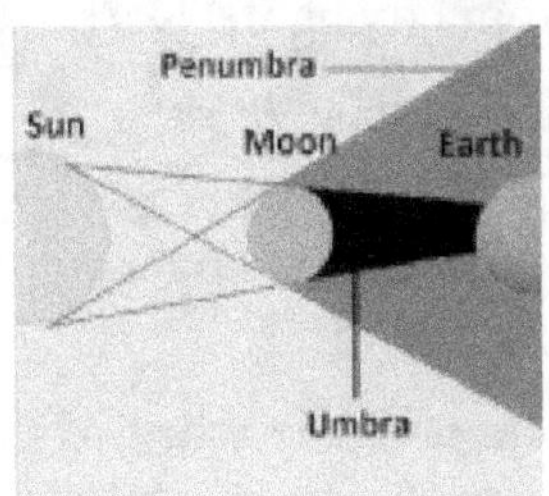

Earth casts two distinct shadows: the penumbra, a partial outer shadow, and the umbra, a complete inner shadow. The lunar eclipse experience unfolds as the Moon enters these shadowy realms. In the penumbral phase, a subtle shading occurs, often challenging the observer's discernment. The penumbra gives way to the more pronounced umbral phase, where the Moon undergoes a remarkable transformation, adopting hues ranging from coppery red to deep brown. This cosmic interaction, a celestial play of shadows, invites us to contemplate the profound cosmic relationships that shape our nocturnal skies.

Tips for Observing Lunar Eclipses

1. Check the Date and Time: Find out when the lunar eclipse will occur in your location. Plan your observation time accordingly, as lunar eclipses can be visible for several hours.
2. Weather Conditions: Ensure clear skies for optimal viewing. Check the weather forecast in advance to choose a location with minimal cloud cover.
3. Moon's Path: Determine the Moon's path across the sky during the eclipse. This helps you find the best vantage point for observation.
4. Use Binoculars or a Telescope: While a lunar eclipse is visible with the naked eye, using binoculars or a telescope enhances the details, allowing you to see the Moon's surface features and the changing colors more clearly.
5. Bring Warm Clothing: Lunar eclipses can be lengthy events, especially during totality. Dress warmly, especially if you're planning to spend an

extended period outdoors.

6. Choose a Dark Location: To observe subtle details and enjoy the full beauty of a lunar eclipse, pick a location away from city lights. Dark skies provide a better view of the Moon's changing appearance.

7. Set Up a Comfortable Viewing Area: Bring a blanket or reclining chair for comfortable viewing. Lunar eclipses unfold slowly, so having a relaxed setup enhances the experience.

8. Capture the Moment: Consider bringing a camera or smartphone to capture photographs of the lunar eclipse. Use a tripod to ensure stable shots during longer exposures.

9. Observe the Entire Event: A lunar eclipse has distinct phases, including the penumbral phase and various stages of partial and total eclipse. Be patient and observe the entire event to witness the Moon's changing appearance.

10. Enjoy the Experience: Take a moment to appreciate the celestial spectacle. Share the experience with friends or family, and marvel at the beauty of our cosmic dance in the night sky.

Equipment List for Observing a Lunar Eclipse

1. For observing the a Lunar Eclipse, you don't need sophisticated equipment. Binoculars can enhance your experience by revealing more details on the lunar surface.

2. If you have access to a telescope, even a small one, it can provide a closer look at craters and lunar mountains.

3. A lunar filter for your telescope is also beneficial to reduce the Moon's brightness, allowing you to see more subtle details.

4. Even without any equipment, the Moon is visible to the naked eye, making it an excellent target for stargazers of all levels.

Tips for Lunar Observation

The moon has captivated human beings for centuries with its mystical beauty and mesmerizing presence in the night sky. Whether you are a new amateur astronomer, a teacher looking to introduce your students to the wonders of the universe, or a curious individual eager to explore the realms of astronomy, lunar observation is an excellent starting point. In this subchapter, we will provide you with essential tips and guidelines to enhance your lunar observation experience.

1. Timing is Everything: The moon's appearance changes throughout the month, with different phases and angles of illumination. To observe the moon in its full glory, choose a clear night when the moon is in a waxing or waning gibbous phase. These phases offer the best visibility and allow you to observe the moon's craters, mountains, and maria (dark, flat areas).

2. Equipment Essentials: While you can observe the moon with just your eyes, a pair of binoculars or a telescope will greatly enhance your experience. For beginners, a small telescope with a decent aperture is recommended. It will allow you to see the moon's features in more detail, such as the craters and ridges.

3. Learn the Lunar Geography: Familiarize yourself with the moon's geography before your observation session. Use a lunar map or smartphone app to identify notable features like the Sea of Tranquility and Tycho crater. Understanding the moon's topography will make your observations more meaningful and exciting.

4. Experiment with Different Magnifications: When using a telescope, try different eyepieces to adjust the magnification. Higher magnification will bring small features into focus, while lower magnification will allow you to observe larger areas. Experimenting with different magnifications will help you find the best balance between detail and overall view.

5. Patience and Steadiness: Lunar observation requires patience and a steady hand. The moon may appear to be moving quickly due to its rotation, so it's important to adjust your telescope or binoculars to compensate for its movement. Take your time and allow your eyes to adjust to the darkness for a more immersive experience.

6. Capture the Moment: Consider keeping a lunar observation journal or taking photographs to document your progress and discoveries. This will not

only serve as a personal record but also allow you to track changes in the moon's appearance over time.

7. Utilize the many online resources available, free of charge, including NASA's Daily Moon Guide and Moon Trek, NASA's interactive guide to the Moon.

Remember, lunar observation is just the beginning of your astronomical journey. As you become more comfortable with the moon, consider exploring other celestial objects such as planets, stars, and galaxies. The wonders of the universe await you, so keep gazing at the stars and always stay curious.

Earth's Tidal Dance - Lunar Influence on Tides

The moon's gravitational influence on Earth extends beyond its visual presence in the night sky. The interplay of gravitational forces between Earth and the moon gives rise to tides—a celestial dance that shapes coastlines and ecosystems. Understanding the lunar connection to tides adds a dynamic layer to our exploration of Earth-moon interactions.

The tidal forces exerted by the moon have led to a state of synchronous rotation, where the same side of the moon always faces Earth. This tidal resonance becomes a celestial waltz, echoing the harmonious dance of Earth and its cosmic companion through the ages.

Human Footprints and Robotic Pioneers

The moon has been a tangible destination for human exploration. From the iconic Apollo missions to the first steps taken by astronauts on the lunar surface, human footprints have left an indelible mark on the cosmic soil. The chapter explores the scientific and cultural significance of these historic lunar missions.

Beyond human exploration, robotic missions continue to unveil lunar secrets. From orbiters capturing high-resolution images to landers investigating lunar soil, ongoing discoveries contribute to our evolving understanding of the moon's geology and cosmic history.

Unexplained Lunar Anomalies & Mysteries

The moon, despite its seemingly static appearance, harbors transient lunar phenomena—unexplained flashes of light, color changes, and other anomalies. These mysteries, observed by astronomers over centuries, add an element of intrigue to our exploration of the moon's enigmatic features.

Unusual geological formations, such as the Reiner Gamma swirls, present lunar features that challenge conventional explanations. Investigating these anomalies becomes a cosmic detective story, prompting scientists to unravel the mysteries concealed within the moon's ancient terrain. The origin of Reiner Gamma—like other lunar swirls—is not completely understood. It is associated with a localized magnetic field, but not with any particular irregularities in the surface. Similar features have been discovered in Mare Ingenii and Mare Marginis by orbiting spacecraft.

Earthrise and Cosmic Perspectives

The Apollo missions gifted humanity with iconic images of Earthrise from the lunar surface. These moments captured the fragility and beauty of our home planet against the stark lunar landscape, fostering a profound shift in cosmic perspectives and environmental awareness.

Earthrise becomes a symbol of interconnectedness, inspiring philosophical reflections on the human experience and our place in the vast cosmic tapestry, transforming our perspectives on the collective consciousness of humanity.

Future Lunar Exploration - Beyond the Horizon

Looking to the future, the Artemis program aims to return humans to the moon, fostering international collaboration and paving the way for sustainable lunar exploration. The Lunar Gateway, a proposed space station orbiting the moon, becomes a stepping stone for future cosmic endeavors.

Future lunar exploration holds the promise of scientific endeavors, from studying lunar water ice to investigating cosmic mysteries. As humanity prepares to venture beyond the horizon once again, the moon stands as a celestial frontier, inviting us to unlock its remaining secrets and continue our cosmic odyssey.

In this exploration of Moonlit Marvels, we have navigated the ever-changing phases and intricate features of Earth's cosmic companion. From the celestial ballet of lunar phases to the cosmic mosaic of craters and maria, the moon has revealed its cosmic secrets. Cultural interpretations, the lunar influence on tides, human and robotic exploration, lunar anomalies, Earthrise moments, and future lunar endeavors enrich our understanding of the moon's profound impact on both the scientific and philosophical dimensions of our cosmic journey.

Chapter 6: The Majestic Dance of the Planets

In the cosmic ballet, planets perform a mesmerizing dance. Let's join this celestial choreography and witness the enchanting movements of our planetary neighbors.

The Planetary Symphony Begins

The cosmic overture commences as the planets take their positions on the celestial stage. Like dancers in a grand ballet, these celestial spheres engage in an intricate choreography that unfolds across the vast canvas of the night sky. The majesty of their movements, both predictable and awe-inspiring, invites us to participate in the planetary symphony that has played out for eons.

In this chapter, we will take you on an awe-inspiring journey through our very own Solar System. Whether you are a new amateur astronomer, a curious kid, an adult venturing into astronomy, an entry-level telescope purchaser, or a teacher looking to inspire your students, this subchapter is designed to provide you with a beginner's guide to the wonders of the Solar System.

The Solar System is a vast expanse of celestial bodies, all orbiting around our central star, the Sun. It is home to eight planets, countless moons, asteroids, comets, and other intriguing objects. Our journey begins by exploring the Sun, the powerhouse that provides us with light and heat. Discover its structure, the incredible fusion reactions happening within, and its profound influence on the Solar System.

In the grand cosmic ballet, planets engage in a mesmerizing dance, unfolding a celestial symphony that has captivated humanity for eons. This planetary choreography, both predictable and awe-inspiring, invites us to explore the wonders of our Solar System. As we embark on this cosmic journey, we delve into the Sun's profound influence. Our exploration then takes us through the diverse planets, moons, asteroids, and comets that populate our celestial neighborhood.

The inner planets, Mercury, Venus, Earth, and Mars, reveal their unique characteristics and offer a celestial spectacle accessible even to novice stargazers. Armed with a telescope, observers can witness Mercury's fleeting ballet, Venus's radiant glow, Earth's lunar dance, and Mars's intriguing surface features. The outer giants, Jupiter, Saturn, Uranus, and Neptune, become celestial gems under the lens, showcasing their intricate details and captivating ring systems.

The planetary dance intensifies with alignments and conjunctions, creating celestial formations of unparalleled beauty. This cosmic choreography unfolds along the ecliptic, a cosmic stage for planetary movements that has fascinated cultures throughout history. Planetary retrogrades, though optical illusions, add drama to Mars's performance, highlighting the complexity of the cosmic ballet.

The Moon, Earth's celestial partner, becomes an integral part of the planetary dance through lunar occultations and dynamic partnerships with other planets. Concluding our journey, we emphasize the importance of the right equipment for observers, offering guidance to entry-level telescope purchasers and encouraging teachers to inspire the next generation of astronomers.

As we conclude our exploration of the planetary dance, we recognize its timeless nature—a testament to the enduring beauty and harmony of our celestial home. Join in the marvel of the planets' majesty, and through their eternal dance, find a reflection of the cosmic wonders that have inspired humanity across the ages. So, join us on this celestial adventure as we dive into the depths of the Solar System, unravel its secrets, and discover the wonders that await us in the night sky.

Next, we will embark on a tour of the planets, starting with the closest neighbor to the Sun, Mercury. Learn about the unique characteristics of each planet, from Venus with its thick atmosphere, to Mars with its mysterious red surface, and the gas giants, Jupiter and Saturn, with their mesmerizing rings. Delve into the icy giants, Uranus and Neptune, and their enigmatic moons, before venturing to the outer reaches of the Solar System where the dwarf planet Pluto resides.

But the Solar System is not limited to planets alone. We will also explore the fascinating moons that orbit around these celestial bodies. From Earth's very own natural satellite, the Moon, to Jupiter's captivating moon, Europa, believed to potentially harbor life beneath its icy surface, each moon has its own unique story to tell.

Furthermore, we will unravel the mysteries of asteroids and comets, the remnants of the early Solar System, and their role in shaping the planets we know today. Discover the captivating stories of famous comets like Halley's Comet, which grace our skies only once in a lifetime.

Throughout this section, we will also provide you with tips, guides, and resources to enhance your stargazing experience. From understanding the best times to observe different planets, to finding the right telescope for your needs, we will equip you with the tools to explore the Solar System with confidence.

So, join us on this celestial adventure as we dive into the depths of the Solar System, unravel its secrets, and discover the wonders that await us in the night sky.

The Dazzling Inner-Planet Wanderers

We will now explore the inner planets of our solar system and learn how to observe them with your telescope. Whether you are a new amateur astronomer, a curious kid, or an adult just getting started with stargazing, this guide is perfect for you.

The inner planets, also known as the terrestrial planets, include Mercury, Venus, Earth, and Mars. Unlike the outer gas giants, these planets are primarily composed of solid materials. Observing these celestial bodies can be an exciting and rewarding experience, as they are relatively close to Earth and can be seen with a modest telescope.

STARRY NIGHTS: A BEGINNER'S JOURNEY INTO ASTRONOMY

The night sky holds a myriad of wonders waiting to be discovered. As you embark on your stargazing adventures, remember to be patient, persistent, and open to the awe-inspiring beauty that lies just beyond our atmosphere. Let the stars guide you, and may the magic of the night sky forever illuminate your journey into the cosmos. Once you have your telescope ready, it's time to explore the inner planets.

A good starting point is a refractor or reflector telescope with a moderate aperture, around 4 to 6 inches. These telescopes offer a balance between affordability and performance, making them ideal for beginners. Once you have your telescope ready, it's time to explore the inner planets.

- Mercury, the swiftest of the planets, performs a fleet-footed ballet as it dances between the Sun and the horizon. Its brief appearances during dawn and dusk offer fleeting glimpses of its graceful choreography against the canvas of sunrise or sunset. Mercury is the closest planet to the Sun, and can be challenging to observe due to its proximity to the Sun's glare. Look for it during twilight, just before sunrise or after sunset, when the sky is relatively dark.

- Venus, on the other hand, is often referred to as the "evening star" or "morning star" due to its brightness. It is easily visible shortly after sunset or before sunrise. Venus graces the celestial stage with elegance. Its brilliance, unmatched by any other celestial body, bathes the night sky in a soft glow. As the "sister planet" to Earth, Venus becomes a cosmic companion in our journey through the planetary dance.

- Earth, our home planet, engages in a delicate orbital pas de deux with the Moon, creating a mesmerizing celestial waltz. This dance, witnessed by stargazers throughout history, weaves a narrative of phases and eclipses against the cosmic backdrop. This dance, witnessed by stargazers throughout history, weaves a narrative of phases and eclipses against the cosmic backdrop.

- Mars, often called the "Red Planet," offers an intriguing sight with its reddish hue. Mars, the red planet, exhibits seasonal changes and polar ice caps observable through telescopes. The shifting patterns of frost and ice reveal the Martian climate's dynamic nature, while features like Olympus Mons and Valles Marineris showcase the planet's geological diversity. Telescopic exploration captures Martian dust storms sweeping across the planet's surface. Observations of the Martian atmosphere provide valuable data for understanding the planet's weather patterns and atmospheric dynamics.

Teachers can utilize this subchapter to introduce kids to the wonders of astronomy. Engaging students in stargazing activities, such as observing the inner planets, can spark their curiosity and foster a love for science. Provide them with simple sky charts and encourage them to keep a journal of their observations, noting the date, time, and any interesting details they observe.

Observing the inner planets is an excellent starting point for new amateur astronomers, kids, and adults new to astronomy. By following the tips and guides provided in this subchapter, you will be well on your way to exploring the wonders of our solar system. So grab your telescope, head outside, and let the journey begin!

Observing the Outer Planets

One of the most exciting aspects of astronomy is the opportunity to observe the outer planets of our solar system. These giant gas giants, known as Jupiter, Saturn, Uranus, and Neptune, offer a fascinating glimpse into the vastness and diversity of our cosmic neighborhood. In this subchapter, we will explore the techniques and tools necessary to observe these distant worlds, providing tips and resources for new amateur astronomers, kids and adults new to astronomy, entry-level telescope purchasers, and teachers.

When it comes to observing the outer planets, a telescope is your greatest ally. Even a small telescope can reveal incredible details on Jupiter and Saturn. It's important to invest in a telescope with good optics and a solid mount, as this will greatly enhance your viewing experience. Additionally, consider purchasing a planetary eyepiece, which is specifically designed for observing the planets, allowing for high magnification and crisp views.

- Jupiter, the largest planet in our solar system, is a favorite among stargazers. With its distinct cloud bands and the iconic Great Red Spot, it never fails to impress. Look for Jupiter in the eastern sky after sunset, and you'll be rewarded with stunning views of its colorful atmosphere and its four largest moons, known as the Galilean moons. These moons appear as tiny points of light surrounding the planet. Telescopic observation transforms Jupiter into a vibrant world of atmospheric dynamics. The Great Red Spot, a swirling storm larger than Earth, captivates observers. Jupiter's Galilean moons—Io, Europa, Ganymede, and Callisto—dance in cosmic harmony, revealing their unique characteristics. Telescopes capture the drama of volcanic activity on its surface. Io's ever-changing landscape, marked by volcanic plumes and eruptions, becomes a testament to the dynamic forces at play in our solar system.

- Saturn, with its magnificent ring system, is another celestial gem. Look for it in the southeastern sky in the late evening hours. Through a telescope, you'll be able to see the rings and even some of Saturn's larger moons. It's a truly awe-inspiring sight that never fails to captivate. Saturn's majestic rings, viewed through a telescope, unfold in unprecedented detail. The enigmatic hexagon-shaped storm at Saturn's north pole becomes a celestial mystery, inviting astronomers to explore the dynamics of this atmospheric phenomenon. Saturn's moon Enceladus, with its icy exterior, conceals a subsurface ocean. Telescopic observation reveals the plumes of water vapor erupting from Enceladus, hinting at the potential for liquid water beneath its frozen shell—a tantalizing prospect for astrobiology.

- Uranus and Neptune, though smaller and fainter, are still worth seeking out. They can be a bit more challenging to locate, but with a star chart or a smartphone app, you'll be able to pinpoint their locations. While they may appear as small blue-green disks, the knowledge that you're observing these distant worlds will fill you with a sense of wonder and awe.

Planetary Alignments and Conjunctions

The planets, in their cosmic dance, align with precision, creating celestial formations that captivate the observer. Alignments such as oppositions and quadratures showcase the harmonious movements of the planets within the solar system.

Conjunctions, where planets appear close together in the night sky, offer celestial encounters of unparalleled beauty. Witnessing the close proximity of Jupiter and Saturn or the meeting of Venus and Mars becomes a moment of cosmic poetry.

Cosmic Stage for the Planetary Dance

The planetary dance unfolds along the ecliptic, an imaginary plane defined by Earth's orbit around the Sun. This cosmic stage becomes the theater for the majestic performances of the planets. Observing the planets along the ecliptic provides insights into the dynamics of our solar system and the interconnected nature of planetary movements. Throughout history, cultures around the world have observed and documented the planetary dance. Ancient astronomers, devoid of modern technology, meticulously recorded the positions and movements of the planets, laying the foundation for our understanding of celestial mechanics. In the modern era, space exploration has provided unprecedented insights into the planetary dance. Robotic missions to Mars,

Jupiter, and Saturn have transformed our perception of these distant worlds, allowing us to witness their beauty up close and unravel the mysteries of their celestial choreography.

The dance of planets extends beyond the celestial realm, influencing Earth's climate, tides, and even human affairs. The gravitational interactions between planets create subtle yet profound effects that echo throughout our interconnected solar system. While modern science emphasizes the physical dynamics of the planetary dance, astrology has long associated the positions of planets with metaphysical significance. Exploring the historical and cultural aspects of planetary symbolism adds depth to our understanding of the cosmic ballet.

Mars, in its intricate ballet, performs a cosmic waltz in reverse during its retrograde motion. This apparent backward dance, caused by the differing orbital speeds of Earth and Mars, adds a touch of celestial drama to the Red Planet's performance.

Planetary retrogrades, though an optical illusion, become celestial phenomena that captivate astronomers and stargazers alike. Understanding the mechanics behind retrograde motion deepens our appreciation for the complexity of the planetary dance.

Observing the inner planets is an excellent starting point for new amateur astronomers, kids, and adults new to astronomy. By following the tips and guides provided, you will be well on your way to exploring the wonders of our solar system. Observing the outer planets is a thrilling adventure in stargazing. With a good telescope, patience, and a sense of curiosity, you can explore the beauty and complexity of Jupiter, Saturn, Uranus, and Neptune. These distant worlds offer a glimpse into the wonders of our solar system and beyond, inspiring both kids and adults alike. So grab your telescope, head outside on a clear night, and prepare to be amazed by the wonders of the outer planets. So grab your telescope, head outside, and let the journey begin!

With a good telescope, patience, and a sense of curiosity, you can explore the beauty and complexity of Jupiter, Saturn, Uranus, and Neptune. These distant worlds offer a glimpse into the wonders of our solar system and beyond, inspiring both kids and adults alike. So grab your telescope, head outside on a clear night, and prepare to be amazed by the wonders of the outer planets. Happy stargazing!

Concluding Remarks

To begin your journey, it is essential to have the right equipment. Entry-level telescope purchasers will find this section particularly helpful. A good starting point is a refractor or reflector telescope with a moderate aperture, around 4 to 6 inches. These telescopes offer a balance between affordability and performance, making them ideal for beginners.

Teachers can utilize this subchapter to introduce kids to the wonders of astronomy. Engaging students in stargazing activities, such as observing the inner planets, can spark their curiosity and foster a love for science. Provide them with simple sky charts and encourage them to keep a journal of their observations, noting the date, time, and any interesting details they observe.

As we conclude our exploration of the planetary dance, one undeniable truth emerges—the celestial choreography is eternal. The planets, in their timeless orbits, continue their majestic dance across the cosmic stage. Whether observed through ancient stargazers' eyes or modern telescopic lenses, the planetary dance remains an enduring testament to the beauty and harmony of our celestial home. Join me as we marvel at the majesty of the planets, for in their dance, we find a reflection of the cosmic wonders that have inspired humanity throughout the ages.

Chapter 7: Deep Sky Odyssey: Nebulae, Galaxies, and Star Clusters

Embark on a cosmic odyssey as we journey beyond our solar system. Let's unveil the hidden gems of deep space - nebulae, galaxies, and star clusters that captivate the imagination.

Introduction to Deep-Sky Objects

Welcome to the fascinating world of deep-sky objects! In this chapter, we will explore the wonders that lie beyond our own solar system and delve into the mysteries of distant galaxies, nebulae, and star clusters. Whether you are a new amateur astronomer, a curious kid, or an adult looking to expand your knowledge of the universe, this subchapter is designed to ignite your enthusiasm for stargazing and deepen your understanding of the cosmos.

Deep-sky objects are celestial bodies located outside our solar system and are often observed with the help of telescopes. Unlike planets and stars, which are confined to our Milky Way galaxy, deep-sky objects can be found throughout the vast expanse of the universe. They offer a glimpse into the diversity and beauty of cosmic objects, captivating astronomers and stargazers alike.

One of the most awe-inspiring deep-sky objects is the galaxy. These vast collections of stars, gas, and dust come in various shapes and sizes, ranging from spiral galaxies like our own Milky Way to elliptical and irregular galaxies. Galaxies are the building blocks of the universe, containing billions of stars and harboring countless mysteries waiting to be unraveled.

Nebulae are another captivating type of deep-sky object. These clouds of gas and dust can be found scattered across the night sky, often displaying stunning colors and intricate structures. From the glowing Orion Nebula to the dark and mysterious Horsehead Nebula, each nebula offers a unique and mesmerizing spectacle.

Star clusters, on the other hand, are densely packed groups of stars that formed from the same stellar nursery. They can be classified into two types: open clusters, which contain relatively young stars, and globular clusters, which are composed of much older stars. Observing these clusters allows us to witness the various stages of stellar evolution and gain insights into the life cycles of stars.

As a beginner in astronomy or a new telescope owner, exploring deep-sky objects can be a rewarding experience. Not only do they offer a visual feast for the eyes, but they also provide valuable insights into the vastness and complexity of the universe. Whether you are stargazing from your backyard or venturing out to a dark sky site, this subchapter will equip you with the necessary knowledge and resources to navigate the cosmos and uncover its hidden treasures.

So, grab your telescope, or even a pair of binoculars, and embark on a journey through the depths of space. Join us as we unveil the wonders of deep-sky objects and discover the beauty that lies beyond our own celestial neighborhood. Get ready to be amazed by the breathtaking sights that await you in the vast expanse of the universe.

Star Clusters - Celestial Gatherings of Stellar Jewels

Star clusters are mesmerizing celestial formations that captivate both amateur and professional astronomers alike. These clusters offer a glimpse into the vastness and beauty of our universe, showcasing a multitude of stars grouped together in various patterns and arrangements. Observing star clusters is an excellent way for new amateur astronomers, kids and adults new to astronomy, entry-level telescope purchasers, and teachers to begin their journey into the wonders of the night sky.

One of the most famous star clusters is the Pleiades, also known as the Seven Sisters, located in the constellation of Taurus. This open cluster consists of several young, hot stars, and can be easily observed with the naked eye or binoculars. Pointing your telescope towards the Pleiades will reveal a stunning sight, where the stars appear like diamonds on a velvet black canvas.

Instructions for Observing the Pleiades

- Observing the Pleiades, a stunning open star cluster also known as the Seven Sisters, can be a delightful experience for beginner astronomers. Located in the constellation Taurus, the Pleiades are a prominent celestial object best observed during the winter months in the Northern Hemisphere, particularly from October to March.
- To locate the Pleiades, look toward the eastern sky during the winter season. The distinctive cluster is situated in the constellation Taurus, near the shoulder of the celestial bull. Find a location away from bright city lights, preferably in a dark-sky area, to enhance your chances of seeing this beautiful star cluster clearly.
- Identifying nearby constellations and stars will help you locate Taurus and the Pleiades. The bright orange star Aldebaran is a notable marker in Taurus and can guide you toward the Seven Sisters. The Pleiades are visible to the naked eye, but using binoculars or a small telescope can provide a more detailed and enchanting view.
- Choose a clear, crisp night for observation, as atmospheric conditions significantly impact visibility. Dress warmly, especially during winter months, and allow your eyes time to adjust to the darkness. Using binoculars or a small telescope with low magnification will enhance your experience of observing the Pleiades, revealing the cluster's intricate arrangement of bright stars.
- Enjoy the beauty of the Pleiades, and take note of the distinct blue glow that characterizes this celestial jewel. The simplicity of this observation makes it an ideal starting point for novice astronomers exploring the wonders of the night sky.

Another intriguing cluster is the Beehive Cluster, residing in the constellation of Cancer. This open cluster contains over a thousand stars, and its name is derived from the appearance of a swarm of bees when observed through a telescope. The Beehive Cluster is an excellent target for entry-level telescope purchasers, as it offers a captivating view and can be located with relative ease.

Instructions for Observing the Beehive Cluster

- Observing the Beehive Cluster, also known as M44, can be a rewarding experience for beginner astronomers. Situated in the constellation Cancer, this open star cluster is easily visible to the naked eye and is best observed during the spring months in the Northern Hemisphere.

- To locate the Beehive Cluster, look toward the western sky during the spring season. Cancer is a relatively faint constellation, but the Beehive Cluster is a prominent celestial object within it. Find a location away from bright city lights, preferably in a dark-sky area, to enhance your chances of spotting this beautiful cluster.

- Identify the nearby stars and constellations to navigate to Cancer. The bright stars Castor and Pollux in Gemini can serve as helpful markers, guiding you toward Cancer. Once you locate Cancer, the Beehive Cluster should be visible as a faint fuzzy patch to the naked eye.

- Spring offers clearer skies and comfortable temperatures for stargazing. Choose a clear night for observation and let your eyes adjust to the darkness. While the Beehive Cluster can be seen without optical aids, using binoculars will provide a more detailed and impressive view of the numerous stars in this cluster.

- Appreciate the beauty of the Beehive Cluster, resembling a swarm of bees, and take your time exploring the intricacies of this celestial gathering. The simplicity of this observation makes it an excellent introductory experience for novice astronomers exploring the wonders of the night sky.

Globular clusters, on the other hand, are densely packed spherical formations composed of hundreds of thousands to millions of stars. One of the most prominent examples is the Hercules Cluster, located in the constellation of Hercules. This cluster is visible to the naked eye as a faint fuzzy patch, but observing it through a telescope unveils a breathtaking sight of countless stars tightly packed together.

Instructions for Observing the Hercules Cluster

- Observing the Hercules Cluster, also known as Messier 13 or M13, is a captivating experience for beginner astronomers. Located in the constellation Hercules, this globular star cluster is one of the brightest and most prominent in the night sky. To observe M13, you'll need to wait until the constellation Hercules is well-placed in the night sky, typically during the late spring and summer months in the Northern Hemisphere.

- To find the Hercules Cluster, look for the Hercules constellation, which is characterized by a distinctive keystone shape. During the spring and summer, Hercules is visible high in the night sky, making it an ideal time for observation. The cluster is situated between two bright stars, Eta Herculis and Zeta Herculis, providing useful reference points for locating M13.

- Choose a clear and dark-sky location away from urban light pollution for optimal visibility. While the Hercules Cluster is visible to the naked eye, using binoculars or a small telescope will enhance the viewing experience by revealing the individual stars that make up this dense globular cluster.

- Set up your observing equipment, ensuring it is properly aligned and focused. Locate Hercules in the night sky and center your binoculars or telescope on the region between Eta Herculis and Zeta Herculis. Once centered, you should be able to spot the Hercules Cluster as a fuzzy patch of light. Take your time observing the details and appreciate the dense concentration of stars within this remarkable globular cluster.

- The Hercules Cluster is an excellent celestial object for beginners, offering a fascinating glimpse into the beauty of globular clusters and the wonders of the cosmos.

When observing star clusters, it is important to find a dark location away from city lights, as light pollution can hinder the visibility of these celestial gems. Additionally, using a low magnification eyepiece will provide a wider field of view, allowing you to appreciate the cluster's intricate details and surrounding nebulosity.

For teachers and stargazing beginners, it is worth noting that star clusters can serve as educational tools. By observing different clusters and noting their characteristics, students can learn about stellar evolution, the formation of galaxies, and the vastness of the cosmos.

Observing star clusters is a thrilling experience that opens up a world of wonder and beauty in the night sky. Whether you are a new amateur astronomer, a child or adult exploring astronomy for the first time, an entry-level telescope purchaser, or a teacher looking for educational resources, star clusters provide an excellent starting point for your journey into the cosmos. So grab your telescope, find a dark spot, and prepare to be amazed by the stunning celestial formations that await you.

Observing Nebulae - Cosmic Clouds of Stardust

Nebulae and galaxies are among the most captivating objects in the night sky. These celestial wonders have fascinated astronomers for centuries, and their allure continues to captivate new amateur astronomers, kids, and adults alike. In this subchapter, we will delve into the enchanting realm of nebulae and galaxies, providing essential information for stargazing beginners and entry-level telescope purchasers.

Nebulae, often referred to as "cosmic clouds," are vast regions of gas and dust scattered throughout our galaxy and beyond. These ethereal formations can take on various shapes and colors, ranging from luminous pink and blue hues to intricate wisps resembling celestial paintings. Among the most renowned types are emission nebulae, planetary nebulae, and dark nebulae. Dark nebulae, cosmic shadows cast against glowing regions, create celestial silhouettes that evoke mystery. The Horsehead Nebula, a dark silhouette against the vibrant backdrop of Orion, becomes a cosmic enigma. The exploration of dark nebulae unveils the hidden dimensions of cosmic shadows and the secrets they conceal.

Deep within the cosmic realms, interstellar dust clouds become the cosmic architects of stellar birthplaces. These molecular clouds, draped in cosmic dust, serve as the cradles for new stars. The Orion Molecular Cloud Complex unveils the intricate dance of dust and gas that gives rise to celestial beacons.

Instructions for Observing the Orion Molecular Cloud Complex

- Observing the Orion Molecular Cloud Complex is a rewarding experience for beginner astronomers interested in exploring cosmic nurseries where stars are born. Located in the constellation Orion, this complex includes several nebulae, such as the famous Orion Nebula (Messier 42), Horsehead Nebula, and Flame Nebula. To observe this celestial spectacle, choose a clear night during the winter months when Orion dominates the night sky.

- Orion is a prominent winter constellation and is easily recognizable with its distinctive "belt" formed by the three stars Alnitak, Alnilam, and Mintaka. The Orion Molecular Cloud Complex is situated below Orion's belt, making it a convenient target for observation.

- For the best viewing experience, find a dark-sky location away from city lights to enhance visibility. While some components of the complex, like the Orion Nebula, are visible to the naked eye, using binoculars or a telescope will reveal more intricate details and structures within the nebulae.

- Set up your observing equipment and point it toward the region below Orion's belt. The Orion Nebula is particularly easy to locate and serves as a stunning centerpiece of the complex. Take your time exploring the intricate filaments, bright regions, and dark lanes within the Orion Molecular Cloud Complex. Consider using a narrowband filter to enhance the contrast and visibility of specific features within the nebulae.

- This observation offers a glimpse into the dynamic processes of star formation and the breathtaking beauty of interstellar gas and dust clouds. Enjoy the cosmic wonders of the Orion Molecular Cloud Complex as you delve into the mysteries of stellar birth within this iconic constellation.

Within the heart of molecular clouds, protostars emerge as cosmic infants, enveloped in blankets of gas and dust. The Taurus Molecular Cloud becomes a celestial nursery, witnessing the birth of stars. Exploring the stages of stellar infancy becomes a journey through the cosmic cycles of birth, evolution, and illumination.

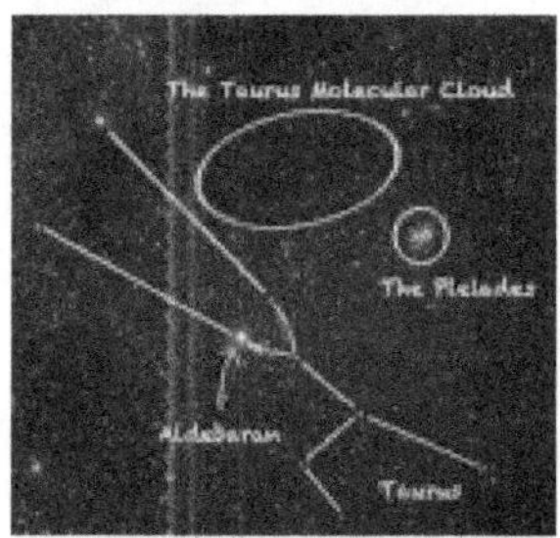

Instructions for Observing the Taurus Molecular Cloud Complex

- Taurus is easily identifiable by the V-shaped cluster of stars known as the Hyades, representing the face of the celestial bull. The Taurus Molecular Cloud Complex extends around the Hyades and encompasses various nebulae and star-forming regions.

- For optimal viewing, head to a dark-sky location away from city lights to enhance visibility. While some components, like the Pleiades, are visible to the naked eye, using binoculars or a telescope will reveal more details within the molecular cloud complex.

- Set up your observing equipment and direct it toward the region around the Hyades cluster. The Crab Nebula and the Pleiades are standout features within the Taurus Molecular Cloud Complex. Take your time exploring the intricate details of the Crab Nebula's filaments and the sparkling stars of the Pleiades. Consider using a narrowband filter to enhance specific features and bring out the contrast in these star-forming regions.

- This observation offers a captivating glimpse into the cosmic processes of stellar birth and the beauty of these molecular clouds. Enjoy the celestial wonders of the Taurus Molecular Cloud Complex as you embark on a journey to uncover the mysteries of star formation within the heart of Taurus.

Dark Nebulae

Dark nebulae appear as opaque, dark patches against the backdrop of stars, obscuring the light emitted by the celestial objects behind them. The famous Horsehead Nebula, located in the constellation Orion, is a prominent dark nebula that resembles a horse's head when observed under the right conditions.

The Horsehead Nebula, a dark silhouette against the vibrant backdrop of Orion, becomes a cosmic enigma. The exploration of dark nebulae unveils the hidden dimensions of cosmic shadows and the secrets they conceal.

Instructions for Observing the Horsehead Nebula

- Observing the Horsehead Nebula is a thrilling endeavor for beginner astronomers, providing a chance to witness one of the most iconic and intricate nebulae in the night sky. Located in the Orion Molecular Cloud Complex, the Horsehead Nebula is situated in the constellation Orion. To observe this cosmic masterpiece, choose a clear winter night when Orion is prominent in the sky.

- Orion is easily recognizable with its distinctive three-star belt. The Horsehead Nebula can be found just below the leftmost star of Orion's belt, known as Alnitak. Alnitak is a bright and massive star, serving as an excellent guidepost for locating the nebula.

- To enhance your viewing experience, head to a dark-sky location away from light pollution. The Horsehead Nebula is a challenging object to observe with the naked eye, so using a telescope equipped with a narrowband filter is recommended. The filter helps isolate specific wavelengths of light emitted by the nebula, making its intricate details more discernible against the backdrop of space.

- Once your telescope is set up, point it toward the region just below Alnitak. The Horsehead Nebula is part of the Orion Molecular Cloud, which also includes the Flame Nebula. Take your time to adjust the focus and observe the intricate silhouette of the Horsehead against the glow of nearby emission nebulae. Patience is key, as careful observation will reveal the distinct shape resembling a horse's head.

- Embark on this celestial journey to witness the ethereal beauty of the Horsehead Nebula, and let the cosmic wonders of the Orion Molecular Cloud Complex unfold before your eyes.

Emission Nebulae

Emission nebulae are known for their vibrant colors, which arise from the ionization of gas by nearby stars. The famous Orion Nebula, situated within the Orion constellation, is a prime example of an emission nebula and a perfect target for beginners. Its distinctive pinkish glow and intricate tendrils are easily visible even with modest telescopes.

Instructions for Observing the Orion Nebula

- Observing the Orion Nebula is a fantastic experience for beginner astronomers, offering a glimpse into a stellar nursery where new stars are born. Located in the constellation Orion, this nebula is easily visible during the winter months. To embark on this celestial journey, choose a clear and crisp winter night when Orion graces the night sky.

- Orion is a prominent and recognizable constellation, often referred to as the Hunter. It features the distinctive three-star belt, and the Orion Nebula can be found just below Orion's belt. Look for a faint glow near the sword of Orion, and you'll be on your way to discovering this cosmic gem.

- No telescope is required to appreciate the beauty of the Orion Nebula; it is visible to the naked eye. However, using binoculars enhances the experience, revealing more details and the cluster of stars within the nebula. If you have a telescope, even a small one, it will provide a closer and more detailed view of the intricate structures within the nebula.

- To optimize your observation, head to a location away from city lights, as light pollution can hinder visibility. Once you've located the Orion Nebula, take a moment to soak in the beauty of this stellar nursery, where young stars emerge from clouds of gas and dust. The Orion Nebula is a captivating celestial sight that showcases the dynamic and creative forces at play in our cosmic neighborhood.

Planetary Nebulae

Planetary nebulae, despite their name, have no connection to planets. These mesmerizing objects are the remnants of dying stars, shedding their outer layers and leaving behind beautiful, glowing shells. The Ring Nebula in the constellation Lyra is a well-known example of a planetary nebula, displaying a distinct ring-like structure that can be observed with a small telescope.

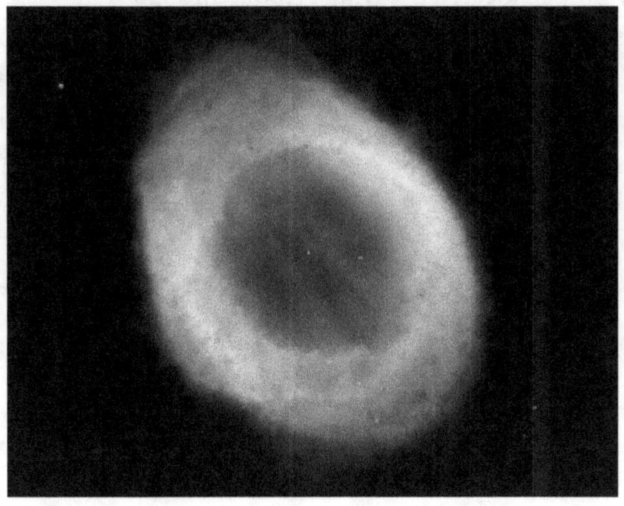

Instructions for Observing the Ring Nebula

- Observing the Ring Nebula is a fascinating endeavor for beginner astronomers, offering a glimpse of a distant planetary nebula. Located in the constellation Lyra, this celestial gem can be observed during the late spring to early autumn months. Lyra is a small but distinctive constellation, and finding the Ring Nebula is an exciting challenge for stargazers.

- To embark on this cosmic journey, wait until the night sky is free from moonlight and head to a location with minimal light pollution. In Lyra, look for the bright star Vega, which is part of the prominent Summer Triangle. The Ring Nebula is situated between the stars Beta Lyrae and Gamma Lyrae, forming a triangular pattern with Vega.

- A small telescope is recommended for observing the Ring Nebula, as it may appear as a faint, smoke-ring-like structure to the naked eye. A telescope with at least a moderate aperture will reveal the distinct ring shape, making the observation more rewarding. Using eyepieces with different magnifications can provide varying levels of detail.

- Once you've located the Ring Nebula through your telescope, take time to appreciate its ethereal beauty. This planetary nebula represents the outer layers of a dying star expelled into space, creating a luminous ring. The Ring Nebula is a captivating celestial object that offers a glimpse into the stellar life cycle, making it a must-see for budding astronomers.

Galactic Marvels - Spirals, Ellipticals, and Irregulars

Galaxies

Galaxies, on the other hand, are immense collections of stars, gas, and dust, each containing billions of stars like our own Milky Way. These cosmic islands come in various shapes and sizes, ranging from spiral galaxies, like the Andromeda Galaxy, to elliptical and irregular galaxies.

Observing galaxies can be an awe-inspiring experience, especially when considering the vast distances they span. While most galaxies are best observed using larger telescopes, some, like the Andromeda Galaxy, can be spotted with the naked eye under dark skies. With the aid of binoculars or a small telescope, you can even witness the spiral arms and core of our neighboring galaxy.

The exploration of spiral galaxies becomes a journey through vast cosmic landscapes, where stars cluster in arms of celestial splendor. Cosmic dust lanes, threading through galaxies, add a layer of complexity to their visual tapestry. The Milky Way's dust lanes, seen from our vantage point, become celestial highways weaving through the galactic landscape. Exploring these cosmic features becomes a journey through the intricate patterns of galactic anatomy. Observing galaxies unveils a captivating celestial panorama, with these immense collections of stars, gas, and dust taking various forms.

The Andromeda Galaxy, a striking spiral galaxy, is a prime example accessible to the naked eye under dark skies, while larger telescopes unveil intricate details of its pinwheel-like arms and core. With binoculars or a small telescope, the cosmic landscapes of spiral galaxies come to life, adorned with celestial splendor and intricate dust lanes resembling cosmic highways. The exploration of galaxies like Andromeda offers a journey through the grand tapestry of galactic anatomy.

Instructions for Observing The Andromeda Galaxy

- Observing the Andromeda Galaxy is an exciting venture for beginner astronomers. Located in the Andromeda constellation, this spiral galaxy can be spotted during the fall and winter months in the Northern Hemisphere. To find Andromeda, look for the distinctive "W" shape of Cassiopeia, and then follow the two longer arms of the "W" towards the horizon. Andromeda is situated near Cassiopeia and appears as a faint, hazy patch to the naked eye under dark skies.

- The optimal time for Andromeda observation is during the fall months, from September to November, when the constellation is high in the night sky. Choose a clear, moonless night for better visibility. Binoculars are a great starting point for observing Andromeda, allowing you to capture the galaxy's larger structure and some of its brighter regions. For more detailed views, consider using a small telescope with low to moderate magnification.

- Identifying the nearby stars in Andromeda can aid in locating the galaxy. Mirach and Mu Andromedae are two prominent stars that can guide you. Mirach, part of the Andromeda constellation, forms one corner of the Great Square of Pegasus and serves as a helpful reference point. Mu Andromedae, also known as Almach, is a colorful double star that adds to the constellation's celestial scenery.

- Set up your observation site away from city lights to reduce light pollution and enhance visibility. Allow your eyes time to adjust to the darkness, and enjoy the journey into the night sky. With a little patience and the right conditions, observing the Andromeda Galaxy can be a rewarding experience for astronomers at any level.

Elliptical galaxies, with their smooth and elongated shapes, exude cosmic elegance. M87, harboring a supermassive black hole in its core, becomes a beacon of celestial mystery. The chapter delves into the timeless dance of stars within elliptical galaxies and the cosmic forces shaping their graceful forms. Elliptical galaxies, characterized by their smooth and elongated shapes, exude cosmic elegance. M87, housing a supermassive black hole, stands as a beacon of celestial mystery, inviting observers to witness the timeless dance of stars within its graceful form. The chapter on elliptical galaxies delves into the cosmic forces shaping these celestial entities, providing a glimpse into the profound interconnectedness of stellar systems.

Instructions for Observing M87

- Observing M87, a fascinating galaxy with a supermassive black hole at its center, can be a captivating experience for beginner astronomers. Located in the Virgo constellation, M87 is best observed during the spring and early summer months in the Northern Hemisphere. Look for the distinctive V-shaped pattern of stars known as the "Spring Triangle" composed of Arcturus, Spica, and Regulus. From there, navigate towards Virgo, and M87 can be found in the region near the star Vindemiatrix.

- The optimal time for M87 observation is during the spring months, from March to June, when the Virgo constellation is well-placed in the night sky. Choose a clear night with minimal light pollution for the best visibility. Binoculars or a small telescope are suitable for observing M87, allowing you to witness its bright core and surrounding details.

- Identifying nearby stars can help you pinpoint M87 in the Virgo constellation. Vindemiatrix, a bright star in Virgo, can serve as a useful reference point. Take note of its position, and then scan the nearby region to locate M87. With the right equipment and a clear sky, you can delve into the intricate details of this captivating galaxy.

- Set up your observation site away from city lights to minimize light pollution, providing a clearer view of the night sky. Allow your eyes some time to adapt to the darkness, enhancing your ability to spot celestial objects. Exploring M87 can offer a beginner astronomer a glimpse into the vast and mysterious realms of our cosmic neighborhood.

Irregular galaxies defy cosmic order with their chaotic shapes and structures. The Magellanic Clouds, satellite galaxies of the Milky Way, embody cosmic anarchy. Exploring irregular galaxies becomes a cosmic journey into the realms of unpredictability, where gravitational interactions sculpt celestial landscapes. Irregular galaxies, defying cosmic order with their chaotic structures, bring an element of unpredictability to the cosmic tableau. The Magellanic Clouds, as satellite galaxies of the Milky Way, embody this cosmic anarchy, and observing them becomes a journey into realms shaped by gravitational interactions. Exploring irregular galaxies unveils celestial landscapes sculpted by cosmic forces, offering a unique perspective on the dynamic and ever-changing nature of our cosmic neighborhood. For the beginning astronomer, the observation of galaxies presents an awe-inspiring opportunity to delve into the diverse and enchanting realms of the universe, with each type offering a unique window into the cosmic ballet of stars, gas, and dust.

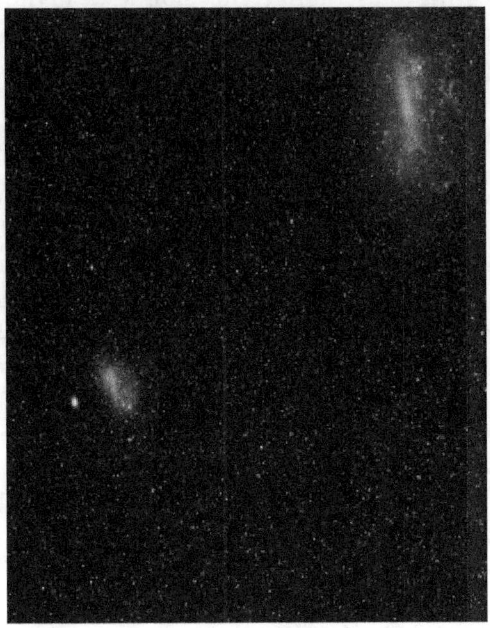

Instructions for Observing the Magellanic Clouds

- Observing the Magellanic Clouds, a pair of irregular galaxies visible from the Southern Hemisphere, can be an exciting experience for beginner astronomers. The Magellanic Clouds, named after the explorer Ferdinand Magellan, consist of the Large Magellanic Cloud (LMC) and the Small Magellanic Cloud (SMC). These galaxies are best observed during the southern hemisphere's summer months, from November to February.

- To locate the Magellanic Clouds, you'll need to be in the Southern Hemisphere and look toward the southern skies. The best time for observation is during the summer months when these galaxies are prominent in the night sky. Find a location away from city lights to minimize light pollution and enhance visibility.

- Identifying nearby constellations and stars will aid in locating the Magellanic Clouds. The constellations Tucana and Hydrus are in the vicinity, serving as useful reference points. Scan the southern sky, and you should be able to spot the two irregular galaxies. Binoculars or a small telescope will enhance your view of the Magellanic Clouds, allowing you to observe their structure and details.

- Ensure you choose a clear night with minimal atmospheric disturbances for optimal viewing. Set up your observation site, and let your eyes adapt to the darkness over some time. Exploring the Magellanic Clouds can provide a fascinating glimpse into the cosmic diversity beyond our own Milky Way for novice astronomers in the Southern Hemisphere.

Cosmic Collisions and Galactic Dynamics

Exploring the phenomenon of cosmic collisions and galactic dynamics opens a window to the awe-inspiring theater of the deep sky. One remarkable example of this celestial drama is found in the Antennae Galaxies. These galaxies serve as a visual testament to the consequences that unfold when galaxies interact and collide. As they come together, gravitational forces give rise to dynamic and transformative events that shape the very fabric of these cosmic islands.

Instructions for Observing the Antennae Galaxies

- Observing the Antennae Galaxies can be a captivating experience for beginner astronomers. These galaxies, also known as NGC 4038 and NGC 4039, are located in the Corvus constellation. To locate them, you can start by identifying Corvus, a small and distinct constellation visible in the southern hemisphere. Look for the distinctive "Y" shape formed by the brightest stars within Corvus, and the Antennae Galaxies are situated near the lower part of this pattern.

- The optimal time for observing the Antennae Galaxies is during the spring months in the southern hemisphere, typically from March to May. During this period, the constellation Corvus is well-placed in the night sky, offering a clear view of the Antennae Galaxies. It's recommended to plan your observation on nights with minimal light pollution for the best visibility.

- For equipment, a telescope with moderate magnification is suitable for observing the Antennae Galaxies. A telescope with an aperture of at least 4 inches (100 mm) will reveal some details of the galactic structures. Additionally, using a low-power eyepiece will provide a wider field of view, enhancing the observation of the interacting galaxy pair. Consider bringing along a star chart or a stargazing app to help you identify Corvus and locate the precise position of the Antennae Galaxies in the night sky.

- Once you've set up your telescope and identified Corvus, aim your telescope at the coordinates of NGC 4038 and NGC 4039. Take your time to observe the galactic interactions and the unique features of these colliding galaxies. Patience and clear skies are key to a rewarding experience, allowing you to appreciate the cosmic dance unfolding in the Antennae Galaxies as they interact and shape the celestial landscape.

Tidal forces, a consequence of gravitational interactions during galactic mergers, play a pivotal role in sculpting the intricate forms that emerge from these celestial unions. The Whirlpool Galaxy and its companion offer a striking illustration of the tidal forces at play during such interactions. This captivating dance of galaxies is a cosmic narrative that unfolds on a grand scale, where the gravitational embrace leads to the creation of new structures and the rearrangement of stellar material.

Witnessing galactic mergers is akin to observing celestial unions that result in profound transformations. These cosmic dances, driven by the powerful forces of gravity, leave an indelible mark on the structures and dynamics of galaxies. It is through these interactions that galaxies undergo metamorphoses, evolving into new configurations that stand as a testament to the enduring dynamism of our cosmic neighborhood. Studying these collisions not only enriches our understanding of galactic evolution but also provides a glimpse into the intricate interplay of cosmic forces shaping the vast tapestry of the universe. As a beginning astronomer, delving into the captivating world of cosmic collisions and galactic dynamics promises a journey filled with celestial wonders and profound revelations about the interconnectedness of the cosmos.

Cosmic Scale and the Limits of Exploration

Embarking on the journey into the cosmos, astronomers face the daunting challenges posed by the vast scales of the universe. Cosmic distances, measured in astronomical units, light-years, and parsecs, serve as the yardsticks for our exploration. An astronomical unit, roughly the distance from the Earth to the

Sun, is a fundamental measure in our solar system. However, as we venture beyond, light-years and parsecs become indispensable units, illustrating the expansive reaches of interstellar and intergalactic space.

Telescopes act as our cosmic vessels, enabling us to traverse these mind-boggling distances. The Hubble Space Telescope, an iconic example, has been a pivotal tool in unveiling the secrets of the universe. Its sharp gaze has penetrated the veils of space, capturing images that transport us to the farthest cosmic frontiers. The advent of next-generation observatories promises even greater capabilities, propelling us toward a deeper understanding of the cosmic tapestry.

The tools of cosmic exploration extend beyond the visible spectrum, delving into infrared, radio, and other wavelengths. These technologies offer unique insights into celestial phenomena, unveiling cosmic secrets hidden from the human eye. As we navigate the limitless expanses, telescopic exploration becomes a testament to human ingenuity, pushing the boundaries of our perception and knowledge.

Yet, the chapter prompts contemplation on the inherent limits of our exploration. The finite speed of light imposes constraints on the real-time observation of distant cosmic events. What we observe today may have occurred millions or even billions of years ago, emphasizing the time-traveling nature of astronomical observations. Despite our advanced instruments, there are cosmic realms beyond our current reach, leaving tantalizing mysteries waiting to be unraveled.

The chapter leaves astronomers with reflections on the ongoing quest to push the boundaries of knowledge in the infinite expanse of deep space. The cosmic scale, with its awe-inspiring dimensions, challenges us to continuously innovate and evolve our tools of exploration. The pursuit of understanding the cosmos remains an eternal journey, urging astronomers to seek answers to the profound questions that linger in the vastness of the universe.

Concluding Remarks

In this cosmic odyssey, we have unveiled the hidden gems of deep space—nebulae, galaxies, and star clusters that captivate the imagination. From the ethereal beauty of emission and reflection nebulae to the cosmic elegance of galaxies and the celestial gatherings of star clusters, the deep sky has unfolded its cosmic wonders. Cosmic dust, galactic dynamics, dark nebulae, and extragalactic phenomena have enriched our exploration of the universe's profound mysteries. As we navigate the vast scales of cosmic distances and embrace the limits of exploration, the deep sky continues to beckon, inviting us to embark on an ever-expanding cosmic journey.

For those new to stargazing and contemplating the purchase of their first telescope, observing nebulae and galaxies can be a rewarding endeavor. These celestial objects provide breathtaking views and offer an opportunity to explore the wonders of the universe from the comfort of your own backyard.

Whether you are a teacher seeking to inspire young minds or an individual embarking on a personal journey into astronomy, observing nebulae and galaxies will undoubtedly leave you in awe of the vastness and beauty of the cosmos. So grab your telescope, venture out into the night, and let the wonders of nebulae and galaxies unfold before your eyes.

Chapter 8: Capturing the Night Sky

Introduction to Astrophotography

Astrophotography, the art of capturing stunning images of celestial objects, is a captivating aspect of astronomy that allows us to bring the wonders of the universe closer to home. In this subchapter, we will delve into the exciting world of astrophotography, providing you with the knowledge and tools to explore the cosmos through the lens of your camera.

Whether you are a new amateur astronomer, a teacher looking to engage students with the wonders of the night sky, or an entry-level telescope purchaser, this subchapter is designed to introduce you to the basics of astrophotography. Even kids and adults new to astronomy will find valuable information here to begin their journey into the captivating realm of astrophotography.

First and foremost, we will cover the essential equipment needed for beginning astrophotography. From cameras and lenses to tripods and mounts, we will guide you through the process of selecting the right gear for your astrophotography adventures. We will also discuss the importance of dark skies and how to find the perfect location to capture awe-inspiring images.

Next, we will explore the various techniques and settings necessary for successful astrophotography. You will learn about long-exposure photography, stacking images, and post-processing techniques to enhance your captured images. We will also provide tips for focusing on celestial objects, dealing with light pollution, and capturing breathtaking landscapes with the night sky as your backdrop.

Throughout this subchapter, we will include practical exercises and step-by-step guides to help you apply the concepts discussed. From photographing the moon and planets to capturing deep-sky objects such as nebulae and galaxies, you will gain the skills necessary to capture stunning images of the cosmos.

Furthermore, we will share valuable resources, websites, and online communities where you can continue to learn and connect with fellow astrophotographers. We believe that sharing experiences and knowledge is an integral part of this hobby, and we encourage you to join the vibrant community of astrophotography enthusiasts.

So, whether you aspire to capture the mesmerizing beauty of the Milky Way or simply want to document your stargazing experiences, this subchapter will equip you with the fundamentals of astrophotography. Get ready to embark on a journey that merges art and science, as you capture the wonders of the night sky in breathtaking detail.

Choosing the Right Camera and Equipment

In the vast world of astronomy, one of the most rewarding experiences is capturing the beauty of the night sky through photography. Whether you are a new amateur astronomer, a teacher wanting to inspire young minds, or simply someone who is fascinated by the wonders of the universe, selecting the right camera and equipment is crucial to capturing those breathtaking moments.

When it comes to choosing a camera, there are several factors to consider. Firstly, determine your budget and level of expertise. For beginners, a point-and-shoot camera or a smartphone with a good camera can be a great starting point. These options are affordable and relatively easy to use, allowing you to experiment and learn the ropes of astrophotography.

If you're ready to take your skills to the next level, consider investing in a DSLR camera. DSLRs offer greater control over settings, such as exposure time and ISO, allowing you to capture more detailed and stunning images of celestial objects. Additionally, DSLRs often have interchangeable lenses, which can further enhance your photography capabilities. Wide-angle lenses are ideal for capturing expansive views of the night sky, while telephoto lenses can help you zoom in on specific celestial objects like the moon or planets.

Aside from the camera, there are other essential equipment to consider. A sturdy tripod is a must-have for long-exposure astrophotography, as it helps eliminate camera shake and ensures sharp images. Additionally, investing in a remote shutter release or intervalometer can further reduce camera shake and allow for longer exposures.

To enhance your observing experience, consider investing in a telescope or binoculars. When choosing a telescope, factors such as aperture size, optical quality, and portability should be taken into account. Beginners may find a refractor or a Dobsonian telescope more user-friendly, while more advanced astronomers might prefer a reflector or a compound telescope.

Lastly, don't forget to invest in a good star chart or smartphone app to help you navigate the night sky. These resources can assist you in identifying constellations, planets, and other celestial objects, making your stargazing experience even more rewarding.

Remember, choosing the right camera and equipment is just the beginning of your journey into astrophotography. Practice, patience, and a willingness to learn are key to capturing stunning images of the night sky. So go ahead, embrace the wonders of the universe and let your creativity shine through your lens.

Happy stargazing!

Tips for Beginner Astrophotographers

Astrophotography is a captivating and rewarding hobby that allows you to capture the beauty of the night sky. Whether you are a new amateur astronomer, a teacher looking to inspire young minds, or an entry-level telescope purchaser eager to explore the wonders of the universe, these tips will help you embark on your journey as a beginner astrophotographer.

1. Start with simple equipment: As a beginner, there is no need to invest in expensive and complex equipment right away. Begin with a DSLR camera and a sturdy tripod. Familiarize yourself with the basics of astrophotography before diving into more advanced gear.

2. Learn the night sky: Understanding the night sky is crucial for successful astrophotography. Familiarize yourself with the constellations, stars, and planets visible in your location. Use smartphone apps or star charts to identify celestial objects and plan your shoots.

3. Choose the right location: Light pollution can significantly impact the quality of your astrophotos. Seek out dark sky locations away from city lights to capture clear and vivid images. National parks or remote rural areas are ideal for astrophotography.

4. Master the basics of camera settings: Experiment with exposure times, ISO, and aperture settings to find the right balance for capturing stars and deep-sky objects. Start with shorter exposures to avoid star trails and gradually increase the exposure time as you gain experience.

5. Use a remote shutter release or timer: To avoid camera shake, use a remote shutter release or the built-in timer function on your camera. This will ensure sharp and crisp images.

6. Experiment with different techniques: Astrophotography offers a multitude of creative possibilities. Try different techniques such as long-exposure star trails, time-lapse videos, or wide-field shots of the Milky Way. Don't be afraid to experiment and find your unique style.

7. Practice image processing: Post-processing is an essential part of astrophotography. Learn basic editing techniques to enhance your images and bring out the details you captured. Software like Adobe Photoshop or specialized astrophotography programs can help you refine your photos.

8. Connect with the astrophotography community: Join online forums, social media groups, or local astronomy clubs to connect with fellow astrophotographers. Sharing your work, asking for advice, and learning from experienced astrophotographers will accelerate your progress.

Remember, astrophotography is a journey that requires patience and dedication. Be prepared to learn from your mistakes and embrace the process.

With time, practice, and the right mindset, you will capture breathtaking images of the cosmos and embark on a truly awe-inspiring adventure.

Chapter 9: Stargazing Etiquette and Safety

Respecting Dark Sky Areas

One of the most important aspects of stargazing and astronomy is the appreciation and preservation of dark sky areas. Dark sky areas are regions with minimal light pollution, allowing for optimal viewing of celestial objects. In these areas, the night sky comes alive, revealing countless stars, galaxies, and other celestial wonders. As astronomers, it is our responsibility to respect and protect these precious dark sky areas for future generations.

First and foremost, it is crucial to understand the concept of light pollution. Light pollution refers to the excessive or misdirected artificial light that brightens the night sky, making it difficult to observe stars and other celestial objects. It is primarily caused by streetlights, buildings, and other sources of urban lighting. By being mindful of our own light usage, we can contribute to reducing light pollution and preserving the darkness of the night sky.

To respect dark sky areas, there are several practices that every stargazer should follow. Firstly, when visiting a dark sky area, always make sure to use red lights instead of white lights. Red lights have a longer wavelength and are less disruptive to one's night vision, allowing for better observation of the night sky. Avoid using flashlights or any bright lights that can disturb other stargazers.

Furthermore, it is essential to leave no trace when visiting dark sky areas. Always dispose of your waste properly and pack out everything you bring in. Avoid leaving any litter or disturbing the natural environment. By practicing Leave No Trace principles, we can ensure that dark sky areas remain pristine and protected.

In addition, it is crucial to be respectful of fellow stargazers and their equipment. Avoid shining lights directly into telescopes or disrupting others' observations. Remember that everyone is there to enjoy the beauty of the night sky, so maintain a quiet and peaceful atmosphere.

Finally, spread awareness about the importance of dark sky preservation. Educate others about light pollution and its effects on our ability to appreciate the wonders of the universe. Encourage your friends, family, and community to make small changes to reduce light pollution, such as using shielded outdoor lighting and turning off unnecessary lights at night.

By respecting dark sky areas, we not only enhance our own stargazing experiences but also contribute to the preservation of our natural world. Let us all be responsible stewards of the night sky and work together to ensure that future generations can continue to marvel at the beauty of the universe.

Protecting Your Eyes and Equipment

As you embark on your exciting journey into the world of astronomy, it is crucial to ensure the safety of both your eyes and your equipment. Stargazing is a mesmerizing experience, but it can also pose risks if proper precautions are not taken. In this subchapter, we will explore essential tips and guidelines to protect your eyes and equipment while exploring the magnificent night sky.

First and foremost, let's discuss the importance of safeguarding your eyes. Staring directly at the sun or other bright celestial objects without adequate protection can cause permanent damage to your vision. Therefore, whenever observing the sun, it is crucial to use certified solar filters specifically designed for telescopes or binoculars. These filters will block harmful ultraviolet and infrared rays, allowing you to safely observe our closest star.

Furthermore, when stargazing at night, it is advisable to give your eyes time to adjust to the darkness. Avoid looking at bright lights or using your smartphone excessively, as this can disrupt your night vision. Red LED flashlights are recommended for reading star charts and adjusting your equipment, as they have a minimal impact on your night vision.

In addition to protecting your eyes, it is vital to take care of your equipment. Telescopes and binoculars are delicate instruments that require proper handling to ensure their longevity. Always transport and store your equipment in a protective case to prevent any accidental damage. When setting up your telescope, make sure it is placed on a stable and level surface to avoid any unnecessary vibrations.

To safeguard your equipment from the elements, invest in quality telescope covers or protective enclosures. These will shield your telescope from dust, moisture, and temperature fluctuations, ensuring optimal performance and longevity.

Another crucial aspect of protecting your equipment is cleaning it properly. Dust and dirt can accumulate on the lenses and mirrors, affecting the image quality. Use a gentle brush or compressed air to remove loose particles, and if necessary, use a specialty cleaning solution and microfiber cloth to wipe the surfaces carefully.

By following these guidelines, you will not only safeguard your eyes and equipment but also enhance your overall stargazing experience. Remember, safety should always be a top priority when exploring the wonders of the night sky.

Whether you are a new amateur astronomer, a parent guiding your curious child, or a teacher introducing astronomy to your students, these tips will help you navigate the awe-inspiring world of stargazing with confidence and care.

Happy stargazing!

Dealing with Weather Conditions

When it comes to stargazing, weather conditions play a significant role in determining the quality of your observation. As a beginner in astronomy, understanding and learning how to deal with various weather conditions is crucial for maximizing your stargazing experience. In this chapter, we will explore the different weather conditions you may encounter and provide you with tips and strategies to overcome any challenges they may present.

1. Clear Skies: Clear skies are every stargazer's dream. With no clouds present, you can fully immerse yourself in the wonders of the night sky. However, even on clear nights, factors like light pollution, humidity, and atmospheric

instability can affect the visibility of celestial objects. We will discuss how to choose the best observing locations, manage light pollution, and use filters to enhance your stargazing experience.

2. Cloudy Nights: Cloudy nights can be disappointing, especially when you have planned a stargazing session. However, it doesn't mean you have to give up on astronomy entirely. We will discuss alternative activities such as learning about constellations, studying star charts, or exploring the science behind celestial objects. Additionally, we will provide tips on how to predict weather patterns and plan your stargazing sessions accordingly.

3. Rain and Thunderstorms: Rain and thunderstorms are not favorable conditions for stargazing. We will discuss how to protect your telescope and equipment during such weather events and provide suggestions for engaging indoor activities like reading astronomy books or watching documentaries to continue your astronomical journey.

4. Cold Weather: Stargazing during colder months can be challenging, but it can also provide unique opportunities. We will discuss how to dress appropriately for cold weather, prepare telescopes for low temperatures, and provide tips for staying warm and comfortable during long observation sessions.

5. Windy Conditions: Wind can affect the stability of telescopes, causing vibrations that can blur your view. We will provide tips on how to stabilize your telescope and suggest techniques for minimizing the impact of wind on your observations.

By understanding how to deal with different weather conditions, you can adapt and make the most out of your stargazing experiences. Whether it's finding alternative activities, protecting your equipment, or choosing the right location, this chapter will equip you with the necessary knowledge to overcome any obstacles and fully enjoy the wonders of the night sky. Remember, patience and persistence are key to becoming an experienced amateur astronomer, and learning to adapt to various weather conditions will only enhance your skills and appreciation for the celestial wonders above.

Chapter 10: Resources for Further Exploration

Recommended Books and Websites

One of the most exciting aspects of embarking on a journey into astronomy is the wealth of resources available to enhance your learning and stargazing experience. Whether you are a new amateur astronomer, a kid or adult new to astronomy, an entry-level telescope purchaser, or a teacher seeking to inspire the next generation, this subchapter is dedicated to providing you with a comprehensive list of recommended books and websites that will serve as valuable tools in your astronomical pursuits.

Books have long been a trusted source of knowledge and inspiration, and there are numerous titles that cater specifically to stargazing beginners. "Astronomy for Dummies" by Stephen P. Maran is an excellent starting point, offering a comprehensive guide to understanding the universe, constellations, and celestial objects. "NightWatch: A Practical Guide to Viewing the Universe" by Terence Dickinson is another highly recommended book, providing practical tips, star charts, and stunning images to aid your stargazing adventures.

For those seeking a more interactive learning experience, websites can be a treasure trove of information. One such website is "Sky & Telescope," a leading resource for astronomy enthusiasts. Their website offers a wealth of articles, guides, and forums where you can connect with fellow stargazers and seek expert advice. Another fantastic online resource is "NASA's Astronomy Picture of the Day," which showcases breathtaking images of the universe along with detailed explanations.

If you are interested in exploring the night sky with your children or students, "The Kids Book of the Night Sky" by Ann Love and Jane Drake is an engaging and educational read. It introduces young readers to the wonders of the night sky through captivating stories, activities, and fun facts. Additionally, the website "KidsAstronomy.com" provides interactive lessons, games, and quizzes designed specifically for young astronomy enthusiasts.

Remember, the key to a successful stargazing journey lies in equipping yourself with the right knowledge and resources. These recommended books and websites have been carefully curated to ensure that you have access to the best tools available. So grab your telescope, immerse yourself in the wonders of the cosmos, and let these resources guide you on your astronomical adventure.

Joining Astronomy Clubs and Organizations

One of the most exciting aspects of starting your journey into astronomy is the opportunity to connect with like-minded enthusiasts who share your passion for the night sky. Joining astronomy clubs and organizations not only provides a supportive community but also opens up a world of knowledge, resources, and experiences that will enhance your stargazing adventures.

Astronomy clubs are formed by individuals who are enthusiastic about astronomy and are dedicated to promoting and advancing the science and hobby. These clubs often organize regular meetings, star parties, and observing sessions, where members can interact, learn from experienced astronomers, and share their own observations. It's a great way to meet new friends, exchange knowledge, and even collaborate on research projects.

For new amateur astronomers, joining an astronomy club can be particularly beneficial. Experienced members are usually more than willing to offer guidance, advice, and recommendations on telescopes, binoculars, and other stargazing equipment. They can help you navigate the complex world of astronomy catalogs, star charts, and observing techniques. Additionally, many clubs organize workshops and lectures on various astronomy topics, providing valuable educational opportunities.

Checkout the Astronomical League[1] to find an astronomy club in your state!

Kids and adults new to astronomy will find that joining an astronomy club can be an exciting and enriching experience. Most clubs organize outreach programs, where members visit schools, community centers, and other venues to share their knowledge and passion for astronomy. These programs often

---

1. https://www.astroleague.org/astronomy-clubs-usa-state/

include telescope viewing sessions, interactive presentations, and hands-on activities that inspire and engage young minds. For teachers, astronomy clubs can be a valuable resource, providing access to educational materials, guest speakers, and field trips that complement their curriculum.

When looking for an astronomy club to join, consider factors such as location, meeting frequency, and the specific interests of the club. Some clubs focus on visual observing, while others specialize in astrophotography or specific areas of research. It's important to find a club that aligns with your interests and goals.

In conclusion, joining an astronomy club or organization is highly recommended for new amateur astronomers, kids, adults, entry-level telescope purchasers, and teachers. The camaraderie, support, and wealth of knowledge that these clubs provide are invaluable for individuals starting their journey into astronomy. Whether you're looking to make new friends, learn from experienced astronomers, or share your own observations, joining an astronomy club will undoubtedly enhance your stargazing experience. So, don't hesitate to reach out and become part of this vibrant and exciting community. The night sky awaits!

Attending Stargazing Events

One of the most exciting aspects of being an amateur astronomer is attending stargazing events. These gatherings offer a unique opportunity to connect with fellow enthusiasts, learn from experienced astronomers, and explore the wonders of the night sky together. Whether you are a kid or an adult new to astronomy, attending stargazing events can greatly enhance your understanding and appreciation of the cosmos.

Stargazing events come in various forms, from local club meetings to organized star parties held at dark sky locations. These gatherings provide a safe and supportive environment for beginners to learn and ask questions about astronomy. Attending these events is a fantastic way to expand your knowledge, gain hands-on experience with telescopes, and meet like-minded individuals who share your passion for the stars.

STARRY NIGHTS: A BEGINNER'S JOURNEY INTO ASTRONOMY

When attending a stargazing event, it is helpful to come prepared. First and foremost, dress appropriately for the weather. Even on warm nights, it can get chilly when you're outside for extended periods. Consider bringing a jacket, hat, and gloves to stay comfortable. If you're attending an outdoor star party, don't forget to bring blankets or chairs for seating, as well as insect repellent to fend off any unwanted critters.

Another essential item to bring is a red flashlight. Unlike regular flashlights, red lights have a longer wavelength, which allows your eyes to maintain their sensitivity to low light. This is crucial for stargazing since it takes about 20-30 minutes for your eyes to fully adapt to the darkness. Using a red flashlight will ensure you can read star charts and navigate without compromising your night vision or disturbing others.

At stargazing events, you will have the opportunity to observe celestial objects through different types of telescopes. If you are a new telescope purchaser, attending these events can help you make an informed decision about which telescope best suits your needs. Experienced astronomers are often more than happy to provide guidance and share their expertise, making it easier for beginners to navigate the sometimes overwhelming world of telescopes.

In addition to telescope viewing, many stargazing events offer informative talks and workshops. These sessions cover a wide range of topics, from basic stargazing techniques to astrophotography tips. Take advantage of these opportunities to expand your knowledge and learn from seasoned astronomers. Don't be afraid to ask questions and engage in discussions – everyone at these events is there because they share a common interest and love for the cosmos.

Overall, attending stargazing events is an excellent way to immerse yourself in the world of astronomy. Whether you are a beginner or a seasoned enthusiast, these gatherings offer a chance to connect with others, learn new skills, and marvel at the beauty of the night sky. So pack your telescope, grab your red flashlight, and get ready to embark on a captivating journey into the cosmos at your next stargazing event!

Chapter 11: Teaching Astronomy to Others

Tips for Teaching Astronomy to Kids

Astronomy is a fascinating subject that can captivate the imagination of both kids and adults alike. Teaching astronomy to kids can be a rewarding experience, as it not only instills a love for science but also encourages curiosity about the universe. In this subchapter, we will explore some tips and techniques for effectively teaching astronomy to kids, ensuring an engaging and educational experience.

1. Start with the Basics: Begin by introducing kids to the basic concepts of astronomy, such as the planets, stars, and the solar system. Use simple language and visual aids to help them understand these abstract concepts easily.

2. Hands-On Activities: Incorporate hands-on activities to make learning astronomy more interactive and enjoyable. Set up a mini planetarium or create models of the solar system to help kids visualize and understand the scale and distances in space.

3. Use Visuals and Multimedia: Utilize visual aids such as pictures, videos, and interactive online resources to make astronomy more visually appealing and comprehensible. Show them images of galaxies, nebulas, and other celestial objects to stimulate their curiosity.

4. Night Sky Observations: Organize stargazing sessions to give kids a chance to observe the night sky through a telescope. Teach them how to identify constellations, planets, and other celestial objects. Encourage them to keep a journal to record their observations and discoveries.

5. Field Trips: Plan visits to planetariums, observatories, or science museums that have astronomy exhibits. These trips can provide kids with a hands-on experience and expose them to more advanced astronomical concepts.

6. Storytelling: Make astronomy come alive by incorporating storytelling into your lessons. Share fascinating stories about the constellations and their

mythological origins. This can add an element of wonder and imagination to their learning experience.

7. Encourage Questions and Exploration: Create a safe and open space for kids to ask questions and explore their curiosities about the universe. Encourage them to conduct their own research or participate in citizen science projects to foster independent learning.

8. Collaboration and Group Projects: Engage kids in group projects that involve building models of the solar system, creating posters, or organizing astronomy-themed events. This promotes teamwork and allows them to share their knowledge with others.

Teaching astronomy to kids is all about nurturing their curiosity and creating a sense of wonder about the universe. By following these tips, you can inspire a lifelong interest in astronomy and help them develop a deeper understanding of the cosmos. So, grab your telescope and embark on an exciting journey into the mysteries of the universe with your young astronomers!

Incorporating Astronomy into the Classroom

One of the most effective ways to foster an interest in astronomy among new amateur astronomers, kids, and adults alike is through the incorporation of astronomy into the classroom. By bringing the wonders of the universe into the educational setting, teachers have the unique opportunity to inspire and engage students with the mysteries of the cosmos.

Introducing astronomy in the classroom can be a powerful tool for capturing the attention of young minds. By utilizing interactive teaching methods, teachers can make the subject come alive, igniting a curiosity and passion for the stars. Hands-on activities such as creating models of the solar system, observing celestial events, and conducting experiments can help solidify concepts and make them more relatable.

Teachers can also utilize various resources to enhance their astronomy lessons. Books like "Starry Nights: A Beginner's Journey into Astronomy" can serve as excellent guides, providing comprehensive information about the night sky, constellations, and celestial objects. Additionally, online platforms and mobile applications offer a wealth of interactive tools, including virtual sky maps, space exploration simulators, and access to real-time astronomical data.

Incorporating astronomy into the curriculum can also promote interdisciplinary learning. By integrating astronomy with subjects like physics, mathematics, and history, teachers can provide a holistic approach to education. For example, students can explore the laws of planetary motion while studying the life and contributions of famous astronomers throughout history. This multidisciplinary approach not only enhances students' understanding of astronomy but also develops critical thinking and problem-solving skills.

Furthermore, astronomy can inspire creativity and imagination. Encouraging students to write stories or poems about the stars, create artwork based on astronomical phenomena, or even compose and perform songs can foster a deep appreciation for the beauty and vastness of the universe.

For entry-level telescope purchasers, classroom integration can be particularly valuable. By offering guidance on telescope selection, usage, and maintenance, teachers can empower their students to explore the night sky beyond the classroom walls. Organizing stargazing events or astronomy clubs can further enhance the learning experience, encouraging collaboration, and providing opportunities for practical observations.

By incorporating astronomy into the classroom, teachers can not only nurture a love for stargazing among their students but also inspire future generations of astronomers. Whether through interactive lessons, interdisciplinary learning, or practical observations, the wonders of the cosmos can become an integral part of the educational journey, captivating the minds of both kids and adults alike.

Inspiring the Next Generation of Stargazers

As the night sky twinkles above us, it holds countless mysteries waiting to be unraveled. The beauty and vastness of the universe have captured the imaginations of generations past and continue to inspire new amateur astronomers, both young and old. In this subchapter, "Inspiring the Next Generation of Stargazers," we delve into the captivating world of stargazing, offering guidance and resources for those embarking on their celestial journey.

For kids and adults new to astronomy, the night sky can seem overwhelming at first. However, with the right guidance, it can become an enchanting playground of discovery. We believe that nurturing an interest in stargazing from an early age is crucial to fostering a lifelong passion. This subchapter aims to provide entry points into the world of astronomy, making it accessible and engaging for beginners of all ages.

One of the most important aspects of inspiring the next generation of stargazers is education. We understand the need for clear and concise explanations, free from complex jargon. With this in mind, we have compiled a range of tips, guides, and resources tailored specifically for beginners. From understanding the different types of telescopes available to learning how to navigate the night sky using star charts, our aim is to make the learning process enjoyable and rewarding.

Teachers play a vital role in introducing astronomy to young minds. This subchapter offers valuable insights and lesson plans to help educators incorporate stargazing into their curriculum. We provide suggestions for hands-on activities, such as creating homemade telescopes or conducting star-gazing field trips. By engaging students in observational astronomy, we hope to ignite their curiosity and encourage further exploration.

Furthermore, for those contemplating the purchase of their first telescope, we offer guidance on choosing the right instrument for their needs and budget. We discuss the various types of telescopes, their features, and their suitability for different observing experiences. Our aim is to empower entry-level telescope purchasers, ensuring they make informed decisions and find the perfect tool to embark on their stargazing adventures.

"Inspiring the Next Generation of Stargazers" is a subchapter dedicated to stargazing for beginners, catering to the needs of new amateur astronomers, kids, adults, entry-level telescope purchasers, and teachers. By providing accessible and engaging content, we hope to instill a love for the cosmos in the hearts of all who embark on this journey. So, grab your telescope, gaze up at the stars, and prepare to be amazed by the wonders of the universe.

Conclusion: Embracing the Magic of the Night Sky

In the vast expanse of the cosmos, our journey through "Starry Nights: A Beginner's Journey Into Astronomy" has been a cosmic odyssey—a tapestry woven with the threads of exploration, wonder, and discovery. As we reach the conclusion of this celestial expedition, let us reflect on the transformative experience that has unfolded under the cosmic canopy.

Our exploration began with the enchanting allure of stargazing—the universal language that transcends borders and connects us to the cosmos. From the myths of ancient civilizations to the modern understanding of celestial wonders, the night sky emerged as an infinite canvas, painting a cosmic symphony of beauty. Stargazing became more than an observation; it became a journey of introspection, inspiration, and connection with the celestial tapestry that has captivated human hearts for millennia.

The chapters unfolded as a guide for aspiring stargazers, empowering them to navigate the night sky with confidence and curiosity. From the basics of stargazing and choosing optimal observation spots to the use of telescopes and the exploration of deep-sky treasures, each chapter provided a roadmap for celestial mastery. The focus on actionable steps and novel ideas aimed not only to impart knowledge but also to inspire a profound connection with the cosmic wonders overhead.

Our journey continued into the realm of stellar stories, where constellations became the living characters of ancient myths and modern tales. From Greek and Hindu mythology to indigenous starlore and cosmic love stories, the night sky transformed into a celestial library of narratives that transcends cultural boundaries. As we explored the stories written in the stars, the connection between humanity and the cosmos became even more tangible, reinforcing the idea that the night sky is a universal canvas upon which we all inscribe our own cosmic stories.

The cosmic odyssey reached its zenith with the unveiling of telescopic marvels, allowing us to peer into the universe with unprecedented clarity. From lunar landscapes to distant galaxies, telescopes transformed our perception, turning distant celestial objects into tangible wonders. We delved into the history of telescopic observation, explored our celestial neighbors, and marveled at the cosmic vistas revealed by the Hubble Space Telescope. The telescope, our cosmic lens, became a gateway to the infinite realms of the cosmos.

As we conclude this cosmic journey, let us celebrate the sense of community that stargazing fosters. Whether in a backyard observatory, at a star party, or through the pages of this celestial guide, the joy of sharing stories, discoveries, and dreams binds us together. I extend my deepest gratitude to each reader who embarked on this cosmic odyssey. Your curiosity and passion for the night sky have fueled this exploration, and I hope the knowledge gained becomes a catalyst for further cosmic adventures.

Our cosmic journey doesn't end here—it merely transforms into a new phase. I encourage each reader to continue exploring the wonders of the universe, armed with newfound skills and an insatiable curiosity. As you gaze upon the night sky, may you find inspiration in every celestial spectacle. Share your stargazing stories, discoveries, and photographs with fellow astronomers, both novice and experienced. In the shared enthusiasm for the cosmos, we find a sense of belonging, fostering a global community of cosmic explorers.

As we come to the end of our journey into the enchanting world of astronomy, it is important to reflect on the incredible experiences and knowledge we have gained. Throughout this book, we have explored the wonders of the night sky, learned about the various celestial objects, and discovered the tools and techniques necessary to embark on our own stargazing adventures. Now, it is time to embrace the magic of the night sky and let it inspire us.

For new amateur astronomers, this book has laid a solid foundation for your astronomical endeavors. You have learned how to navigate the night sky, identify constellations, and understand the movements of the celestial bodies. Armed with this knowledge, you can now confidently explore the cosmos with your telescopes and binoculars.

STARRY NIGHTS: A BEGINNER'S JOURNEY INTO ASTRONOMY

To our young readers, we hope that this book has ignited a passion for the stars within you. Astronomy is a field where curiosity and imagination are celebrated. As you continue to observe the night sky, remember that every twinkle in the darkness holds a story waiting to be discovered. Let your inquisitive minds guide you on a journey through the vastness of space.

For adults new to astronomy, this book has provided you with a gateway to a new and exciting hobby. Whether you choose to stargaze from your backyard or venture to remote locations to witness truly dark skies, the beauty of the universe will captivate you. Remember to embrace the sense of awe and wonder that astronomy offers, allowing it to bring joy and peace to your life.

Entry-level telescope purchasers will find themselves equipped with the necessary knowledge to make informed decisions. You understand the different types of telescopes, their features, and how to maintain them. With your new telescope, the universe becomes your playground, and the possibilities for exploration are endless.

Lastly, to our beloved teachers, we hope that this book has provided you with valuable resources to inspire and educate your students. Astronomy is a subject that sparks curiosity and encourages critical thinking. Use the knowledge gained from this book to create engaging lessons and hands-on activities that will nurture a love for science and the cosmos in your students.

In closing, I express my sincerest hope that the nights ahead are filled with wonder and cosmic discoveries. May the beauty of the night sky continue to inspire, and may each moment spent beneath the celestial dome be a reminder of the vastness, mystery, and enduring beauty of the cosmos. As we bid farewell to this cosmic odyssey, let the stars above be your eternal companions, guiding you through the cosmic tapestry that unfolds in the silent embrace of the night sky.

The night sky holds a myriad of wonders waiting to be discovered. As you embark on your stargazing adventures, remember to be patient, persistent, and open to the awe-inspiring beauty that lies just beyond our atmosphere. Let the stars guide you, and may the magic of the night sky forever illuminate your journey into the cosmos.

Thank you for joining me on this celestial journey. Until we meet again under the cosmic canopy, may your nights be filled with the timeless wonders of the universe.

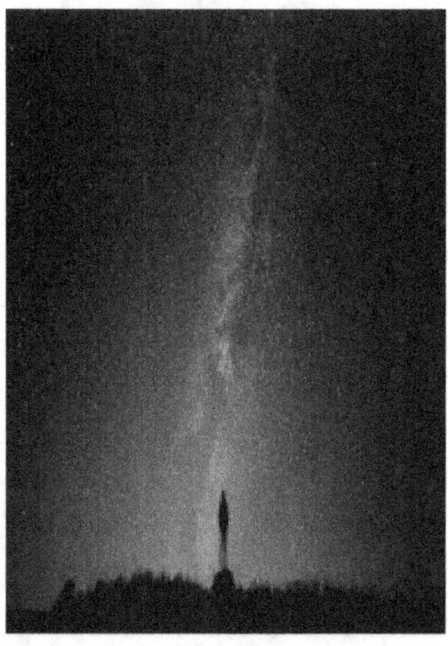

Starry Nights:

A Beginner's Journey into Astronomy

Embark on a celestial journey with 'Starry Nights: A Beginner's Journey into Astronomy,' where knowledge meets passion in an accessible guide designed for beginners. Whether you're navigating the cosmos solo, guiding a child's curiosity, or enhancing classroom lessons, this book promises to illuminate the wonders of the night sky with clarity and enthusiasm. Open the pages and let the cosmic adventure begin.

The author, Larry Culver, is a seasoned aerospace professional, residing in Houston, Texas, who brings a wealth of knowledge and experience to the realm of astronomy. With an impressive background of 35 years in aerospace engineering and software development, Larry has been an integral part of projects with the National Aeronautics and Space Administration (NASA) and the Department of Defense (DoD). His lifelong passion for all things space-related began in his boyhood, shaping a trajectory that intertwined with the exploration of the cosmos.

Retired from a distinguished career in aerospace and business, Larry has dedicated himself to a new mission — the empowerment of young minds through education about the wonders of the universe. His book, "Starry Nights: A Beginner's Journey into Astronomy," reflects not only his expertise but also his commitment to sharing the fascination of the night sky. Larry envisions a world where the next generation is inspired to look up and discover the marvels of the cosmos, fostering a curiosity that transcends boundaries and reaches for the stars.

(N)EVER (S)TOP
LEARNING

Don't miss out!

Visit the website below and you can sign up to receive emails whenever Larry Culver publishes a new book. There's no charge and no obligation.

https://books2read.com/r/B-A-BEPDB-JXVZC

BOOKS2READ

Connecting independent readers to independent writers.

www.ingramcontent.com/pod-product-compliance
Lightning Source LLC
Chambersburg PA
CBHW071317150726
47997CB00002B/509